I0478759

THE WIZARD OF AUTOMATION

This story provides suggestions for developing a new industrial automation system and for starting and running a Process Control System company.

The book offers new ideas for advanced control that can increase plant efficiency, flexibility and reliability.

Although the story is tailored to a process automation system for the oil and gas industry, any automation-oriented business person can relate to the events.

ABSTRACT

Thriving entrepreneurs are fixated by a desire to make a worthwhile achievement, and they follow a business plan that transforms technology to commercial success while minimizing risks.

To be competitive in the marketplace a firm needs to drive innovation in their products and services. And, its technology, while being the core of the company, must be combined with a focus on business strategy and goals.

General professional knowledge and business experience alone are not enough for a company's success. Relevant niche industry knowledge and insight on trends and developments are required.

ABSTRACT

This book offers recommendations that help you overcome the challenges of a Technology Startup or Small & Medium-Sized Enterprise (SME). The example of an automation company is used to present diverse scenarios.

Increase the competitiveness of your company!

Increased process- and factory automation can provide huge benefits to the economic muscle and give businesses competitive advantages and confidence.

The 'hot' technology startup stories often relate to innovations in internet solutions, data communications and other consumer oriented break through – the dream is to be one of the titans of Silicon Valley. While these inventions receive a lot of attention, they seldom impact the efficiency of the process-and manufacturing industries – Oil & Gas, Chemical, Pharmaceutical, Food & Beverage, Glass/Fiber, Power & Utilities, Steel, etc.

Automation has a wide range of applications in many industries. Whether for process automation, manufacturing or solutions for infrastructure tasks, it is a major contributor in helping to increase productivity.

Contents

Contents

Preface

HUH! Who would be interested in a story about the development of a new Automation System?

Non-fiction is great about communicating a business plan with recommendations that can leverage technology. The entity that it misses is the emotional implementation of these recommendations that is at the center of most company activities. The story is a communiqué for relating to the human challenges that are so fundamental to a company. You get to experience the emotional ups and downs alongside the acting persons.

What is inspirational in this story and in real life is not the engineering and entrepreneurial success in the end, but people digging deep within them to overcome seemingly insurmountable hurdles. It is the author's hope that this story makes the readers step back and have a conversation about going into business for themselves and wanting to use technology for the basis of their success.

The story contains two fundamental assets of successful startup firms - confidence and the grasp of real-life business hazards. Indeed, there is a reason for confidence, since if entrepreneurs planned realistically and were cognizant of the business threats; the current condition, that every four out of five startup firms flop, would not occur. The success of startup firms would in fact be high.

Part of the reason this story can seem troubling is because it stops short of far-fetched science fiction. The described invention goes a step further than the technology available today, but generally the book's characters use technology that is not far-fetched. The author's imaginings have, to a great extent come true, in the time since he himself was deeply involved with the detailed definition of an advanced process control system.

While this is not completely a work of fiction, the project and the imagined development efforts are just about factual, but names, characters, places and some events are the author's imagination. The story illustrates that common sense and fixation on accomplishment can inspire people to overcome the frustrations, obstacles, and challenges that come with today's typical changes in a technology company.

Stories about process control systems may be rare, but the author hopes that he can create a personal connection between his audience and his message with this book.

This story encompasses many details to share the author's insights about where professionals in the technology business most often need to focus – its intent is to provide practical examples that can be of help to an entrepreneur or small company manager.

Chapter 1 – OPERATION OF BIG FIRMS

Staring into the screen of the computer monitor, Greg gauged the item of so many recent discussions – the PLC control system. "Is this a real process control system?", he asked himself with his eyes looking at the programmable logic controller (PLC) in the cabinet next to him. He tried to concentrate on encouraging outcomes, but his mind kept wandering to the more troublesome ways this could all play out. He could not erase the potential risks of using a standard function, non-redundant PLC for critical process control from his thoughts since he was exposed to configuration problems with PLC systems on industrial process applications during his last two assignments.

This is the story of two of engineers - Greg Winkler and Robert Cottens - employed by SONARES Engineering. They are testing a process control system.

Before we proceed with the story, permit me to tell you more about Robert and Greg; very different characters. Robert, the lead engineer, is a real go-getter. Always anxious to move the project forward as fast as possible and not hesitate to take credit for any progress, whether or not it was his accomplishment. Greg, on the other hand, is almost a perfectionist who seems to be always looking for better ways of doing things. He is acting as the process control system expert on this job, since he was recently involved with a similar project.

Although Greg preferred to be in the background, he could speak up forcefully when he felt ignored.

Anticipations were high for this new release of software. DDC3, the third-generation beta release was supposedly very advanced. It was described as having all the features of an advanced process control system, which both Greg and Robert have been familiar with, to some extent.

Robert and Greg turned around when they heard the lab door open and watched as Mickey, the representative's sales engineer of the DDC3 software, weaved his way through the labyrinth of instruments filling the room. He carried a coffee cup in each hand and gave one to each of them as he moved into his bench on which a computer for another project was sitting. This was Mickey's daily routine and they appreciated his attitude.

Greg tapped the computer keyboard and an array of colorful images lit up on the monitor and floated in front of him. "What's the decent word?" Robert asked, looking over his shoulder. "I'm just starting my basic control loop analysis. The deviation indication is green and the loop looks normal on every aspect; however, I have not yet simulated a process upset. I should have the preliminary PID profile by the end of the day," Greg responded.

"That's what I want to hear," Robert said, studying the images for a few moments more. He turned back to his booth, sipped his coffee, and let Greg focus on his work. Greg certainly was an experienced control system engineer and had been assigned to this job for final verification of the control system. If things continued to check out, the system would soon be ready for a more comprehensive test drive.

The project timeline was to finish the lab tests and the application configuration in three months and, if rumors were correct, install this beta release in the operations center of the chemical plant complex for a final assessment. If it performed well in that setting for two months, the DDC3 control system would move into full operation. The manufacturer of the DDC3 system had no reliability data, since this was the first system applied in a critical process control application. An unusual situation, considering that the process handled in this chemical plant was from a safety perspective considered critical.

"The library of control functions look good", said Robert as he put his face up close to the monitor and scanned through the function listings to get a more detailed view. He is shortsighted and his nose almost touches the screen, every time he takes his glasses off. "Do you think they will work as described?" he said.

"The functions or the linking of them? Either way, the answer is I sure hope so, since we are here to configure the loops and not to correct any functions," noted Greg as he turned to Robert.

"Don't you have an important presentation today?" asked Greg.

"Yes, it is this afternoon. I present to project management, and then I meet with Bill right after," answered Robert (Bill Orborns was the Project Manager).

"I sure hope you are not going to tell him that everything is perfect since it took us so long to get the system up, and we have not verified a single function until now," replied Greg.

"That is what he expects to hear and I will have to say that things are proceeding well. With all the other problems on this project, I am really the only one who can come up with positive news at this time." Robert responded.

Robert loved this job. The politics of pleasing bosses and project management made it a little less fun, but he thought he was on the verge of something big in his career. It was a great feeling. He reviewed the notes for his talk one last time, and then went for his noontime jog. In spite of the heat, he pushed himself hard. Jogging was his stress relief tool, and with the challenge of a presentation followed by a possible critique by Bill on the slow control system start, he needed the calming effect that his routine provided. He ended his route with a short walk, hands on his hips, while he let his heart rate settle. Then, turning to the changing room next to the restrooms, he cleaned up, and changed into what he thought was a smart-looking suit he brought with him this morning for that presentation.

He reached the conference room, grabbed a cup of water, and slipped into his chair just as Bill Orborns started the meeting. "Good afternoon, everyone, glad you made it on time," Bill said to the assembled project team, while looking at Robert. He moved through some general project items and then shifted to the control system, the main agenda item of this meeting. "I have asked Robert to give us a status update on the control system. You all know that Robert has been leading the configuration and test effort of this system. So Robert, please go ahead. Ah, and I have asked him to be brief, so we will have plenty of time for discussion."

Robert stood at the front of the table and scanned the group. He was pleased to see that everyone's body language was friendly and welcoming. "Hi, everybody." He gave them an uneasy smile as he willed his nerves to settle, then he started his presentation. "Well, we just received the latest software release, DDC3, and after some minor boot-up challenges things appear to be OK. From the library of control functions it looks like they have checked-off the items on the punch-list we submitted three weeks ago. We will start verifying the functions tomorrow. Again, so far so good." He paused and scanned the group to make sure he still had their attention. "It is my experience that control systems are predictable and compliant" he continued. "We have very seldom had reports of unexpected behavior as long as they were used as intended."

"Wait," said Brian Gibson, the senior process engineer. "Has someone used this specific system model not as intended and had an adverse outcome?"

Robert paused, unsure how to answer, and Bill stepped in to rescue him. "Thanks for catching that, Brian. We know of no unreported cases. We do know that the system was being used in a pilot plant application somewhere in the Northeast", Bill added. "Apparently, it had been performing quite well for almost a year."

"What is the system performance record in a process plant, similar to ours?" asked one of the project engineers, who thought that the performance question was important for the discussion. "This is being covered by system supplier direct with our customer" Bill assured him. When Bill sat down, Robert sought to speed things along. Too much focus was given to installation background information.

"As we test this software release, we set our sights on control optimization", Robert said, looking at Brian. "The solutions we came up with are relatively simple and practical to implement. We submitted them to the system supplier, and their analysis indicates that our recommendations are quite easy to incorporate. We are now in the final review period, and the system should be ready for live testing in a couple of months." There were nods from most of the project team members. Then one engineer asked, "You said 'control optimization,' as if we created this. Is this our design, or is it theirs, the customers and the suppliers?"

Robert pasted a smile on his face, but his mind was panicky. Here it was, the topic he wanted to talk with Bill about—but he would not discuss it here. He was, first and foremost, a team player. "My goal is to configure a tool that could then be used for optimization design." It was the best he could come up with on the spot, and he thought it sounded pretty good. "This is how control technology has advanced throughout all of time." That last part was not really relevant, and he hoped he would not be called on it. The meeting then shifted to a commercial focus, and time ran out before any more uncomfortable technical questions could be asked.

Robert was relieved. As Bill closed the meeting, Robert asked him "can I speak with you privately?"

"Nice job back there," said Bill as they walked into his office. "Would you like something to drink? Water? Tea? Coffee?"

"No thanks," Robert said, sitting down at a small table next to Bill's desk. He had left the project meeting satisfied that he had avoided putting Bill on the spot in an open forum. Now that they were alone, he would voice his concerns and get Bill's support for a solution. Bill took a sip as he looked at him. "Your request to speak with me sounded urgent. You haven't been offered another job, have you? Or are you seeking to be transferred to another project?" He was only half joking, always worried about losing key people.

"Nothing like that," Robert said, shaking his head. "This is about our control system reliability. You know that Greg has serious reservations about the systems dependability in our application, and I also have some concerns. As we move closer to going live, they haven't diminished. I am hoping you will have some words of wisdom for me."

He watched him and waited. He knew Bill wouldn't be happy with what he was about to say and sought to buy some time. "Can I have a glass of water?" Bill retrieved a glass of chilled water from his service unit, setting it in front of him as he took his seat again. He did not talk, giving him the opportunity to say his piece.

Robert picked up the glass, held it for a moment, and put it back down without drinking. "Think about it, we are about to install a PLC system that has no proven process control reliability record and all that in a chemical plant. I think that Greg has a valid point about this safety issue and he is on my case for not bringing it to anybody's attention. I am concerned that he may mention it to the customer." Bill remained quiet, and Robert continued "Greg really knows what control reliability issues mean. He experienced problems with a similar system on his last project assignment. And we both know that I initially objected to the selection of this type of system for our project".

Bill folded his arms across his chest. "Wow. You really know how to defend yourself. You and Greg have been testing the system for more than two months and you are telling me now, just before it is scheduled to be installed at site that the control system may not work. Is that what I am hearing?" He furrowed his brow. "I have to admit I am frustrated when I hear you say that you 'initially objected'. You know that way back when the customer selected this system, you said that this technology was not proven but that you were quite confident that we could make it work." He signed quotation marks in the air with his hands as he finished the phrase. "So what's going on? Are you saying it will go scallywag on us and it is not reliable enough to control the plant?"

Bill acted surprised, though Greg had briefly discussed this concern with him at a coffee-break weeks before, which he never mentioned to Robert. "No, at this stage of the game we cannot go to the customer and tell him that all other parts of the project are ready for startup, but the control system may not work." Bill voiced. He paused and frustration was creeping into his voice "Not on my job" he said, pointing his finger at

Robert." You and Greg need to resolve the control system issues more rapidly." Bill stood up from his chair and stared at him. He kept at it until Robert broke eye contact and looked down at the table. "This is your job." His tone was accusatory and Robert blushed. "Let's get it finished in the six or seven weeks," ended Bill. And instead of the "take-it-slow and make sure that everything is solid" Robert was hoping for, Bill was moving in the opposite direction with talk of finishing the system verification completion ahead of the official schedule. Robert said "we will do our best" and left the office.

On the way back to the lab he shook his head as if both to state and deny a personal failing. Just before he reached the entrance door to the lab, he looked at his watch. It was almost 4 p.m. He stopped and decided to turn around and go to his office. He did not want to face Greg and have to tell him what happened in the meeting with Bill. He drifted a minute, thinking about his personal journey to this point and how Greg was openly critical of his handling of the project. He looked at the desk in his office again and took his coat from the hanger. His heart sank. He had come to Bill for solutions and did not get any. Instead he was put under more pressure. Robert decided to go home.

Next day at 7:30 a.m. when Robert entered the lab, Greg was already sitting in front of the monitor and looked up with an unusually serious expression. "What did you tell them?" he said.

"I told them that basically things are OK and that we still had some reliability concerns," said Robert. "What else could I say?"

"Were you able to meet with Bill privately?" Greg asked.

Robert nodded and visibly shuddered when he said, "yeah, and instead of giving us some breathing room on this verification task, he told me that we need to finish all the tests in six weeks." And he found his strength, in spite of Greg's looks of disbelief, to add some assertiveness to his words. "And I still think with the new software release there is a chance that we get there in time."

"Well, considering what happened so far with the past three software releases, I don't know in which world you are living in," Greg replied calmly. He tried hard not to get upset but he wanted to emphasize the reality of the situation.

Greg remained quiet and continued. "Let me tell you what happened yesterday while you were at the meeting. Charlie Gruhn, the Plant Superintendent - you know one of the guys from the customer that you mentioned was involved in this system selection decision – well, he came by with the field maintenance group; you should be glad that you were in the meeting and not here. This was apparently the first time these people from the plant have been told that they are getting a PLC-based process control system" Greg said. "I could not believe what I heard. I was simply amazed".

One techie asked "Hank, you said that we are getting a new type of control system - where are the controllers?" Hank told him that there are no physical controllers with this new type of system anymore. "All resides in this CPU", he said while pointing at the PLC system located in the cabinet. "You will no longer have to maintain so many different devices. Don't worry, you guys will be getting lots of training before we go on-line with this."

Robert interrupted Greg and asked, "Did Hank request you to give them a presentation?" He was concerned that Greg may have made comments to the people that would raise questions about the system's ability to control the plant. "Luckily not," Greg replied. "I had the feeling that he wanted to make the visit as short as possible since he told them that they were already late for a meeting and that they will have to leave."

"So what was the problem?" Robert wanted to know and set down on his chair.

"Well, a few minutes after they left, one of the guys, who told me he was the maintenance supervisor, came back and asked me what I thought about this system and when I felt it may be ready for its field shakedown." Greg responded and continued, "I did not really know what to tell him and said that project from Sonares Engineering was advised that his project team was scheduling the field test in a few months." Greg followed up with "the maintenance team did not seem to have an idea what they are getting. Strange."

Robert, relieved that Greg did not express his concerns during that customer visit, answered, "So, things went relatively smooth. What are you apprehensive about then?"

Greg thought, 'I can't get the message across, even to Robert, and he knows the circumstances. This is a situation where I have been trying for a long time to communicate what I am thinking, but the recipient doesn't want to understand'. It seemed like the harder Greg tried to persuade, the more alienated Robert got. Greg became irritated.

Greg then raised his voice. "In the event that things spin out of control with this PLC system, and in my opinion they will, I feel it is our duty to tell them that we need an automatic backup system" he said.

"Isn't this why we added some analog stations a few months ago?" Robert responded. He was referring to the manual override controllers, added as a fail-guard earlier in the month at Greg's insistence. These manual devices are controlled by simple switches: "Off" - where they would do nothing and the PLC controls would function at full capability, or "On" - where the PLC output signals are manually overwritten.

Greg took a deep breath, then charged. "How often do I need to tell you that I don't think that this will work, even if these manual stations are added? It is simply not adequate to keep the plant under control" Greg said.

"I don't get it. Manual control in case a PLC malfunctions. What's not to work?" Frustration was creeping into Robert's voice. "Okay, suppose I am the operator. I am watching the plant and the PLC controls behavior, I grow concerned and decide to revert to the manual stations," Greg nodded to show that he was following, though he visibly gasped and interrupted Robert.

"You still have not thought through the overall backup control option. We must not only have control redundancy, but most importantly, the proper automatic fallback features in the functions or this will end in a disaster." He added "while this will add quite a significant cost, we really need it."

"Well, the updated function features are supposed to be included in this software release. Why are you getting so upset?" countered Robert.

Greg's tone got angry and his face reddened. "DDC3, third-generation Frankenstein" is the term several people use privately in this lab." He said "you do not even realize their talk had made it also outside the lab walls." And Greg found the strength to add some assertiveness to his words. "And I still think we need to push hard for the full range of capabilities – automatic backup and fallback functions for all complex loops. These are standard features of process control systems for this type of chemical plant application. That is the reason I keep emphasizing this over and over. Of course, the proper solution for this application would have been a fully redundant Automation System, as I stated in the beginning. Unfortunately, PLC suppliers do not offer this solution yet."

"OK, OK, I hear you" Robert countered. "Let's get on with the function test. This constant critique about not having the proper backup control is not helping us complete the configuration of the basic loop functions, which is the main task we are supposed to complete here." Greg is obviously getting on Robert's nerves, but deep inside Robert knows that Greg's points are valid.

And Greg still remembered the panic some of his colleagues showed in those first days during his last start up assignment in the field when the control system malfunctioned and could not end the conversation without having the last word. "Are you telling me that these reliability and application issues are not critical?" he asked Robert. "This is not just squabbling among us," he added.

Greg stood up and paused for a few seconds, and then put his hands into the pockets of his jacket with a sense of finality. "I think we need to bring others in on this. There's too much at stake. But we also need more information before we start sending up flares. These large company politics make communications so complicated."

The whole morning was almost gone and both Robert and Greg got increasingly frustrated with each other's confrontational attitude. Robert was scheduled to go to another meeting regarding control system training issues and left the lab without saying a word to Greg as to why he was leaving.

Greg started the PLC system again and put his feet on the table, taking deep breaths while starring at the monitor screen showing the boot messages. His calm moment was interrupted when two technicians bustled in, sat on the floor behind the PLC cabinet, and opened an access cover to work on something hidden inside. He watched for a few moments, then said, "You seem to be in quite a hurry."

"Oh!" yipped one of the techs, clearly startled. "Sorry, sir, we didn't know anyone was in here."

"What's going on?" said Greg and rose from his chair.

One of the men, holding a piece of equipment in his hands, stood up, while the other continued working. "We are preparing for an addition." He examined the device as he spoke, then bent down and showed his partner. "Watch this connector when you slide it in, and please, be gentle." He stood back up and looked at Greg. 'We are configuring the system for a redundancy upgrade. This entails putting these interconnecting cables in place." Greg said "I know", but he was baffled that the refit was being made so expeditiously.

This was not the kind of news he expected to learn this afternoon. He was delighted and regretted the confrontational position he took just a few minutes ago when talking with Robert. "We are also installing parallel I/O cable assemblies if this change gets approved."

"Well that is even better. Who gave the orders for this?" Greg asked.

"Gosh, sir, we work for a private contractor, and I got these orders from my boss."

Considering that funding and time were in short supply on this project, Greg could hardly get over his surprise of seeing the action on redundancy being taken care of so rapidly. He felt that he was taken seriously again and was almost overjoyed.

With the PLC booted up, Greg began his control function tests again. He configured a simple feedforward loop with an analyzer feedback trim; checked the modes and put in a simulated analyzer signal to verify the loop behavior. The loop acted customary as he manipulated the analyzer feedback within normal range. He then moved the signal beyond the high limit, mimicking an analyzer malfunction. Instead of automatically assuming a fallback value, the control loop 'opened', causing its output to go to zero. "Dammit! The same calamity again" Greg shouted to himself. There was nobody else in the lab. "Will these PLC development engineers ever understand how process control is supposed to work? They don't have the fundamental appreciation of safety and the application know-how required to control a process plant. This is scary!"

Greg spent another two hours randomly going through the punch-list and verifying whether the software malfunctions had been corrected. Most of them had not been taken care of. He was not certain if it was a case of the programmers simply not understanding the functional requirements or the system supplier not assigning a high priority to this project. Anyway, he was very discouraged and as Robert returned from his meeting he said, "We are not making any progress with this software and my bitching and moaning about it to the same people at the supplier did not bring us any results so far. So how can we criticize constructively in a way that leads to the problems being fixed? I don't think that more of the past type of complaining will do any good." And he followed up "I am not sure if it is a case of the software programmers making repeated mistakes or if there is a deeper issue that the DDC3 department does not understand process control. Are we talking to the right people?"

Robert, visibly disturbed, responded "Greg, my meeting about training issues did not go well, can we talk about this tomorrow."

"Of course" Greg said, "I will list the problems I just found and we can discuss the situation tomorrow." Greg could see on Roberts facial expressions that he was under stress and wanted to postpone any discussion.

"What an emotional roller-coaster" whispered Greg to himself, being embarrassed about his outburst about PLC engineers. "Just an hour ago I was thrilled to witness the backup cable installment and now this happens." He felt uneasy at this turn of events. It seemed almost hopeless. It was not the fact that many items on the punch-list had not been corrected that upset him so much. It was that the same situation that happened on his previous project assignment seemed to repeat itself here during the past two months. And within that big engineering company environment with its multi-layered organization, his inability to cope with these problems in a reasonable timeframe was very frustrating and stressful. He vividly remembered the panic situations that these types of system malfunctions caused at the last startup and the scape-goat hunt that followed it.

Greg decided he would return home that afternoon. The physical distance would give him the space and perspective he needed to think through the situation. The large company politics were affecting him and he felt that working with this group would bring down the quality of his work, despite his best efforts. It would also affect his ability to move up in the company. Emotionally off-balance, he started to collect his personal items in preparation for leaving. His half hour ride home was a fog. Recognizing he was tired, he wanted desperately to lighten his load by seeking advice from a trusted friend.

Linda Bowen was his obvious choice. She had worked for a large technology institute in her younger days and then ran her own firm with about a dozen employees for over ten years. She had always been a valuable acquaintance and would listen and give him honest recommendations, he was sure. As he arrived home, he went right to the phone and made the call. Linda picked up. Greg, still upset, said "Hi Linda, sorry to disturb you at this time of the day."

"Well", she responded "ten minutes ago I was still at the store looking at new drapes and you would not have reached me. You sound distressed Greg. Is there something wrong?"

"Yes, things are not going well at the office and the lab. I apologize for bothering you with this."

"Oh, no Greg don't apologize" she replied, "what are friends for? Tell me what happened."

"Well, it is not just this incident. It seems that one thing after another goes wrong with my job at this engineering company. So for almost a year now, I have been exploring a career change"

"Wow, you certainly know how to surprise somebody. How come you haven't mentioned it before?"

"I was going to wait until the difficulties I face on this project are fixed before making moves to leave the company. And even now, both my moral and emotional compass compels me to stay involved until the key issues are resolved; I will need to let my boss know of my plan to terminate the employment so that they have time to familiarize another engineer for that system check-out."

"Do you already have another job lined up?" asked Linda.

"No, but a friend of mine who just joined a startup company, you know Vernon Averoni, has talked to me several times about an engineering position with them."

"But will landing a job at a startup be the right decision for you, an engineer that has worked for a large company for many years? **Discovering what type of environment you work best in is just as important as which company you want to work for**," noted Linda and she followed up with advice. "Before you decide which path is right for you, take some basic things into consideration."

"OK" Greg replied, "I appreciate you helping me."

"My pleasure" said Linda. "We have been friends for so many years, but since we rarely talked about your job, I am not sure how much help I will be."

"I always value your counsel", Greg interjected.

Linda continued "the first thing for you to do is to determine your career goal. Think about where you see yourself in twenty years. Are you hoping to one day become a top-level executive at a large company? Or do you have an entrepreneurial spirit with a passion for starting your own company? While you are thinking about joining Vernon's startup, do not overlook that a larger, more established company will likely offer you better pay and health benefits. But, it may be five years or more before you see a significant salary increase or position move in the company. Startups typically have a flat organizational structure, offering the opportunity to make a significant impact in a shorter time span. Since many startups fail, job security is not the business's biggest asset. And this would be just one of the considerations for you to think about."

Linda then went on to say, "Have you really discovered your work style and passion? Do you thrive in tight-knit teams, frequent collaboration opportunities and a more relaxed company culture? Or do you work best in a more structured corporate environment that provides more of a work-life balance? Take time to explore how you prefer to work to avoid landing in a frustrating job that stunts your personal and professional growth. Think again about your ultimate workplace happiness, including your strengths and weaknesses, and your natural role in a team."

Greg said, "Yes, you are right. There are many things I need to review."

Linda continued, "And make sure you choose the business model that supports your career path. So, what type of position is Vernon's startup company looking to fill? The highest-paying positions are in business development, sales and marketing, not necessarily in engineering. Although the demand for engineers is increasing, so a person like you with solid technical skills may be a hot commodity in the startup space. If you should decide to move, try to make sure that it is profitable" she said.

Linda persisted "Greg, know what you should expect to gain from either a corporate or startup work environment – large vs small company. In a more established company, you will likely work a set schedule, somewhere between the hours of eight to five. As you know, there are typically many layers of managers, directors and top-level executives, so expect to climb the steps of the corporate ladder to earn more pay and notoriety. Large companies are required to adhere to legal compliance, so there may be tasks employees must do, no questions asked. But I think you already know this situation. In a startup environment, expect to work long hours and in small teams across departments. Startups tend to be more transparent, so you will likely know how the company is performing from week to week. You will be required to wear a lot of different hats and may have to take the lead on projects you may perhaps know little about."

Linda then finished by stating "Greg, keep things in perspective. Whether you decide to work for a startup or stay at the larger, more established company, it will have its ups and downs. Again, just remember what your ultimate goal is and what you want to accomplish in your career. It will keep you focused and help you make sound decisions with every opportunity."

"You have given me a lot to think about," Greg said, "I want to hug you with both arms, but unfortunately we are on the phone. It will take me some time to digest it all."

"No worries," she said. "Promise me that you won't overstress yourself and that you take a calm assessment of your prospects." At that point she added "And don't hesitate to call me if you need help. Just make sure that you think everything through before making any moves," then she hung up.

The next day, during his morning jog, Greg used the time to clear his head and weigh his options. He had worked hard at every step of his career, taking the pressure from those above and trying to work with engineers like Robert. As he jogged along, he got more and more convinced that in the large engineering company (although things have been mostly fine in the firm) his contributions that he was so proud of would never be really appreciated. On the other hand, this engineering company treated him so well during all six years he has been with them. The company was not responsible for the technical difficulties he faced during the last couple of months, he says to himself. He felt that his boss should at least be the first in the company to be aware of his intent to leave.

Greg felt nervous and excited, but desperate to get on with a solution to his problem. He put the Belgian waffles and coffee which he regularly fixed for breakfast down in a hurry. Besides, he forgot about his typical coffee refills and could not wait to call Linda again. "Linda", he said, "I think I made up my mind. I am most likely going to resign this job and join the small company. I am going to call my boss and talk to him about it."

"Well, don't let your boss know that you plan to leave in an effort to get more input as you are working through your decision; you would be looking for advice in the wrong place," Linda responded. "Unless you have a very rare boss who is more concerned about your future than about his own or the company's, don't do it. Regard any discussion about your "potential" resignation as equivalent to tendering it. Once you have let the cat out of the bag, it may be near impossible to put it back. Word may get out among your co-workers, and it may affect their attitude about you. Your boss may view what you have revealed as an indication that you will continue looking, even if you don't take this job. And, if you have not yet made a firm decision, all that talk may lead you to make the wrong decision." She continued, "I am a believer in getting advice and insight about a potential job change. But, I think it is dangerous to seek such advice from people whose own jobs and lives will be impacted by your decision."

"Greg, be absolutely sure about the decision to leave before you resign", said Linda. "That may sound obvious, but you should be totally committed to the new job before you resign from your present one. The moment of resignation is not part of the decision process. While your resignation may trigger a counteroffer that you even may accept, a counteroffer should not be the motivating force behind a resignation." She said "besides you may feel uncomfortable about resigning your job, when you worry that your employer will try to talk you out of it. If you can be talked out of your decision, you have not based it on very sound criteria. There may be other criteria that are meaningful to you, but don't miss these basic ones."

"Let me tell you something from my own experience" said Linda "Are you moving to a job where the technology and products represent a positive step for you? Just because the technology is 'leading edge' does not mean it holds value for you. What matters is whether it fits into your plans for your future. Will you be working with people whose attitudes, goals and style are a good match to your own? Are you moving to a business environment where you will be content, if not passionately happy? I may have mentioned all that to you yesterday," Linda said," but it is so important that it is worth repeating."

"Also, will you learn more at this new job than you already know about your work, your business, your industry and yourself? This pertains to both the formal and the informal education you stand to gain. And will you earn what you are worth? Will you be able to earn more as you grow? Also, find out how other people who have quit the company have been treated. Were they walked to the exit by a security guard? Were they treated with respect? Did they get any back pay owed to them? How was unused vacation time handled? While individuals may be handled differently, you will be able to plan better if you know what the typical experience has been. And, never resign unless you have the offer from your friend Vernon in writing. I have seen job offers withdrawn just before, and even after, a new hire shows up for work. It may not be legal, but go fight city hall. I have also seen candidates show up for work only to learn that the position title is not exactly what they were told and that the terms of employment have changed a bit. Avoid surprises; a written offer that includes all the details makes it less likely that there will be serious misunderstandings than an oral offer. A written offer also gives you a legal tool to support your position in a controversy."

"I appreciate all that good advice" said Greg. "Some of it I thought through last night. I could hardly sleep. I was thinking about the resignation letter"

"Well there is certainly a lot to consider when you hand in your resignation", said Linda. "Give your resignation first to your boss, in private; then give a copy to Human Resources. Once you are behind closed doors, all you need to say to your boss is, 'I am sorry to tell you I am resigning my position.' It is not a good idea to let your boss learn of your resignation from someone else, so think twice about who you confide in. The letter should be just one sentence because, sorry to be the cynic, but careers and lives might hinge on this. It can come back and bite you legally. 'I, Greg Winkler, hereby resign my position with SONARES Engineering'. That's it. Sign, seal and deliver. Any other details can be worked out through discussion, including how much notice will be given and when you'll get your last paycheck. If you are forced to take legal action for any reason, or if the company sues you for, say, stealing information, anything you put in your letter can be used against you."

"Also, don't explain. Don't complain. Remember, if you have made a firm decision to take a new job that's right for you, your resignation should not be a discussion about what it would take to get you to stay. That is another matter, and another discussion to have with your boss long before you start pursuing other jobs. Likewise, your resignation should not be a forum where you explain what is wrong with the company. Explanations lead to complaints, and complaints can lead to problems. Even if the Human Resource department wants you to share all your thoughts and feelings in an exit interview, the time of your departure is not the time for you to help fix what's broken at the company. Such meetings can be helpful under the right circumstances, but in my opinion this is no time to open your heart and tell all. It is too risky, and there is zero upside for you. I would politely decline. If your boss or human resources are not aware of any problems that may have led you to resign, they are doing too little and they are asking too late. Anything you say could hurt you. Stride very judiciously. Open up only to those you trust, and only to the extent that you really think it matters."

"Moreover, keep your future to yourself. It seems nice to share your new address with your buddies or your boss. But if, for example, someone thinks your new employer is a competitor, suddenly that comfortable two-week notice can turn into an immediate departure, or worse. If someone presses you, simply say 'I would prefer not to discuss my new job right now.' A headhunter once told me it is the cleanest way to make a transition. 'But, I will certainly call once I get situated because it is important to me to stay in touch with you.'

"And, do not forget to make your last day a good one" added Linda. "When an employer and a worker have a good relationship, which I understand is the case in your situation, parting can be friendly, respectful and cooperative. Once your boss has accepted that you are leaving, let him know that you consider it your responsibility to minimize the difficulty of the transition for the company and for your coworkers. Tell them 'If you would like me to prepare any materials for the person who will replace me, or to actually help train someone, I am ready to do that as long as it does not interfere with a reasonable start date at my new job. How can I help?' That is how you find out how long the company would like your 'notice' to be. Expect and plan on two weeks of some of the hardest work you have done; you owe that to your employer before you leave.

By the way, if you would like a week off between jobs, arrange your start date accordingly. Don't count on the two-week notice being allotted by your present employer. If your boss wants you out immediately, this obviously does not apply. But, you can make the offer anyway. These people have an engineering business to run, and you just told them you no longer want to be a part of it. So, be polite and understanding if you are shown the door."

"Most importantly, said Linda, "Build, do not burn, bridges. Do not dwell on the possible risks and problems that your job change causes. As long as you have calculated carefully, they are transient. What counts is your position in, and contribution to your niche industry as a whole. That is where your true equity in your career lies. No matter who gets upset about the job change you make, the discomfort will pass if your associates are good people. That is why I advocate that the kinds of people you join are as important as the work and the compensation. You will likely see them again. In or out of your presence, the people you work with help define you - they contribute to your knowledge, your philosophy and your reputation. So, add a little extra cement to the foundation of the relationships you have. Build those bridges a little stronger. It's said that 'what goes around, comes around'. I also believe that 'who goes around, comes around'. So no matter how much you want to vent your opinion, swallow your criticisms. Make a point of shaking hands with everyone you have worked with. The nice thing about a handshake is that it does not require words, and it allows you to both emphasize your respect and to hide your negative feelings. Then, prepare to meet your buddies again after you move on'. That's all I can advise you" Linda concluded.

As Greg listened to Linda, he repeatedly said to himself 'this is all common sense', but he does not interrupt her a single time. He recognized that Linda has many years of experience and that her advice to him is sincere. He thanked Linda. "I sure appreciate your advice, Linda. I thank you from the bottom of my heart."

"That's what friends are for. Good luck with your endeavor," said Linda and she hung up.

Greg decided to promptly call his friend Vernon Averoni to confirm that the offer extended to him a few weeks ago was still valid. He knows Vernon well. They have been coworkers for several years until Vernon left the company and cofounded a startup company. Vernon has attempted to convince Greg to join him at his firm for over six months now.

"Hello Vernon" said Greg "how is everything going with you? Is the company still performing well?"

"Good to hear from you Greg. Yeah we are doing great. When will you join us?" responded Vernon.

"Well, that is the reason for my call. Does your offer still stand?" said Greg.

"Am I hearing right? Are you finally considering joining us? That would be so wonderful. Excuse me, but I get all excited. Are you serious?" shouted Vernon.

"Yes I am serious. Can you send me that 'super' employment proposition that we have been talking about last time we met?"

"Will do, right away. This is great! I will have the hiring letter including the attachments forwarded to your home address by DHL; well, even prompter, I will tell my secretary to send the letter immediately to your personal e-mail address," responded Vernon. He followed on "review the package and do not hesitate to contact me at work or at home. You have my cell phone number."

"Thanks Vernon," said Greg and he hung up.

Greg was relieved and enthused; not only did his friend Vernon confirm the employment offer he made two weeks ago, but with Vernon's excited response on the phone Greg felt inspired and confident that his employment with MICGEN Controls would be a great experience to look forward to. Although Greg trusts Vernon, before he accepts the offer, he wanted to evaluate it carefully. He sought to verify that the written proposal truly matched the verbal agreements he discussed with Vernon.

And in less than an hour later, Greg receives the employment offer via e-mail.

Dear Mr. Winkler!

MICGEN Controls Ltd is pleased to formally offer you the position of Engineering Manager.

As discussed, you will be responsible for both the engineering and R&D functions in our company. The duties and responsibilities that you will be expected to carry out are detailed in the attached job description (Enclosure 1). You will report directly to the President, Vernon Averoni. Your starting date will be February 19, 20XX.

Your compensation package includes a weekly salary of $X,XXX payable biweekly, health insurance, life and disability insurance, sick leave, vacation and personal days through our company's employee benefit plan. Please refer to our Employee Benefits Handbook (Enclosure 2) for details.

This job offer is contingent on your passing the mandatory drug screening. This will be arranged once you have acknowledged your acceptance of this job offer.

Please signify your acceptance of this offer by signing and dating this letter where indicated below and signing and dating the standard Confidentiality Agreement (Enclosure 3). These documents should be returned directly to the attention of Vernon Averoni using the business-reply envelope enclosed. A copy of each of these documents is enclosed for your records. We require acceptance by February 2, 20XX.

We look forward to welcoming you to our company. Please let me know if you require any further information. I can be reached directly at (phone number).

Sincerely,
Tina Alexander
Administrative Assistant to Vernon Averoni

Greg thought that it is a good idea to answer back with an email confirming that he had received the written job offer and had signed and sent it back. This way Vernon knows that the employment process is proceeding. Of course, he would not actually forward the e-mail until after the meeting with his boss, Sam Watering. He drafted the response.

Dear Vernon,

I received your formal job offer yesterday evening. I have read through it, signed it and sent it back to you as requested. As suggested, I have kept the second copy.

Thank you again for giving me this exciting opportunity. I look forward to starting employment with MICGEN Controls on February 19, 20XX and to becoming a member of such a dynamic team.

If there is any additional information or paperwork you need please let me know.

Regards,
Greg

Before breakfast, because he did not want to lose last night's train of thought, Greg decided to write the resignation letter. It was, however, not the one-sentence note that Linda recommended, 'I, Greg Winkler, hereby resign my position with SONARES Engineering Inc.'. Greg felt that Sam Watering, his boss, would be offended by such a communication. Of course, Linda was not aware that he and Sam had established a very friendly relationship over the years. So he wrote the following letter:

Dear Sam,

It is with regret that I must inform you of my resignation, to take effect Monday [date], two weeks hereafter.

I will return my keys, access pass, company credit card (cut into two), and laptop so there can be no question of impropriety.

This has been a very difficult personal decision to come to. I have found my time here to be personally

38

and professionally rewarding, and I want to thank you for the manner in which you have treated me throughout our reporting relationship. My personal regard for you, and our positive working relationship both made my decision to leave even more difficult.

I was not unhappy, but was recently approached on behalf of another company and have been offered a role which offers me a tremendous career opportunity I cannot achieve in any other way than by leaving to pursue it, and so I have accepted it.

You may rely on me to conduct myself professionally for the balance of the notice period, and I have prepared a summary of the work I was doing and attached it as well. I would like to speak to you as soon as possible to discuss how you would like me to transition those commitments.

Under the circumstances, I will understand if you wish me to refrain from contact with customers or from attending the office meetings for the duration of my employment. Please advise me as to how you would like me to proceed from this point forward.

Again, my personal thanks to you for the many positive aspects of our relationship and your leadership.

Yours sincerely,
Greg Winkler

Greg fixed his usual Belgian waffles and coffee breakfast and then drove to the office. He was late. On route, he rehearsed the points which Linda advised him and he thought through last night. The objective is to leave the company smoothly, without getting a counter offer. He felt ready for the meeting with Sam.

He had a list of everything he needed to do to a complete and smooth handover written up and ready to go, together with a transition plan. That should make it as easy as possible for Sam to let him go as soon as possible. He identified the best person, in his opinion, to take over for him, in case Sam asked for advice. He planned to pack his personal belongings and the things he wanted from his desk and put them in his car before the meeting started. He had the letter of resignation prepared.

When arriving at the office he intended to immediately try to schedule his meeting with Sam, identify it as a personal issue he needs to discuss, and see if he can schedule it for late afternoon this Friday. This would maximize the time Sam has over the weekend to digest the news and get over it before Greg faces him again on Monday morning. If this Friday is impossible, then he will go for a late-day meeting any other day, for the same reason. Identifying it as a 'personal issue' would most likely also alert Sam to the possibility that something is up and that would help break the news more softly. He will not discuss the termination with Sam on the telephone.

The take away message for Sam needs to be 'He is a great boss. This is a great company, but I have changed and so I have to go. Let's talk about the transition.' But I need to spread that out over a few sentences. My words would be something like: 'Sam, this is a really tough thing to do. I have really enjoyed my time here very much. It has been a privilege to work with you and this company, and I have learned a great deal. I wasn't actually looking for a change, but I have been approached with another opportunity that is completely in line with where I want to take my career. It is simply too good a situation to pass up.

So this envelope contains my letter of resignation, my ID and my company credit card, to take effect Monday morning two weeks from now. I hope you understand. I have really enjoyed working for you. You have taught me a lot and it was a really tough call to make; but in the end, out of deference to my family and myself, I just couldn't say no. I have prepared a list of everything I am working on, and I have an action plan to transition it. Perhaps you can review it, and I will be happy to discuss it and do anything you need me to do to help ease the transition'.

It's Friday and Greg thinks this may actually be the best time to give his notice. He knows that Sam Watering, his boss, is usually relaxed on Fridays, thus he will ask for the termination meeting to be held late-afternoon because it may help everyone involved to avoid the post-meeting awkwardness, and it gives him a couple days to regroup before entering his last two weeks at work.

As he arrived at the office building, the receptionist greeted him with "Robert Cottens has been looking for you Greg." But instead of going to the lab, where he would most likely find Robert, Greg went right away to his office and called Sam. "Sam, I need to discuss a personal issue with you. Would you have time later on this afternoon?"

"Sure," Sam said. "I hope it's nothing serious; let's meet at 4:00 p.m." Greg was relieved that Sam did not ask him any questions on the phone and encouraged that he might actually be able to get through all this before the weekend.

Greg then went to the Lab to go through his test and verification paperwork. He wanted to make sure that the documentation was in order and all test reports were attached, just in case Bill Orborns would want to see a copy on Monday and call a status meeting on the DDC3 verification task. He wanted to be able to provide a history of the complete control system test effort during the past two months. After all, they are supposed to only take the standard IEC 61551 factory acceptance test (FAT) forms – eleven pages of guidelines with about 30 test items – to verify the control system features. This was estimated to be a three week task. It was assumed that all control functions have been proved by the supplier before the FAT started. This was not the case. While system integration and test is considered a fairly unpredictable process, nobody assumed that the basic system functionality has not been verified before start of the FAT.

Greg knew the best he could do now was to put this past two month's test history out of his mind and focus for the next two weeks on testing and retesting the control functions. He set down at the DDC3 monitor, took the last couple of test sheets, started the system and spent the whole day on function tests. He was not interrupted. Robert seems to have vanished. He did not even break for lunch. And, he almost missed the termination meeting time with Sam. It was 3:50 p.m. and before heading up to Sam's office he dated his punch list items and ran to the photocopy machine, which was on the next floor, to make copies for Robert. He put the duplicates on Roberts's desk, walked to the elevator to the 7th floor and was at Sam's office at 4:02 p.m.

Termination meeting

Sam's office door was open, as usual, and Sam was sitting in his chair with his feet on the desk reading what looked like a report. He asked Greg to give him a minute and to sit down. Greg did not sit down. He did not want to invite a lot of conversation by sitting in a relaxed position. He paused and said "Sorry, Boss, this won't take very long, and I don't want to make it any tougher than it already is." Then with the letter in hand, he continued "Boss, after careful deliberation, I have made a commitment to join another organization and will begin working with them effective in two weeks. Please accept this, my letter of resignation. I ask that you take a minute to read my letter before we discuss how we can make my transition as smooth as possible."

Sam's face turned ash as he read the letter. He looked up and said "Greg, you have six years of experience with our company under your belt. You have been one of my ace engineers and you are a sharp designer with a knack for solving problems and getting things done on time. Come on, you can't be serious. What can we do to keep you? Is it the money?"

"Yes, it is a little more money – but that really was not the issue. The issue was the exceptional opportunity fitting to my goals."

"And which company is offering you that opportunity?" Sam asked.

"No, I can't tell you who I have joined – it was a condition of my offer that I not disclose that until I have started. I can share that they are not a competitor."

"Well who are they? Can you not at least tell me what type of company you are going to join?" Sam continued to press, "How can I improve things – please – I want your help"!

"Boss, I am not sure this will be productive. I have a tough time leaving because I like the people here and like working for you, but this is just an outstanding opportunity, one that simply can't be duplicated here. So, as tough as it was, my decision is made. I have given my word, so it's a done deal. I really hope we can focus on the transition," Greg responded.

Sam could see that Greg did not want to change the message and said, "I understand. I accept your resignation and want to work out a smooth transition; let's talk about this Monday." He added "I will let Bill Orborns know about the situation, if he is still in this afternoon, so that you don't have to go through an exit meeting with him. He will be really mad."

"I appreciate that very much Sam" said Greg.

Then, Sam stood up and said, "I wish you all the best in your new job, and in case things don't turn out as anticipated, do not hesitate to call on me."

"Thanks very much" Greg replied. They shook hands, smiled and Greg walked out the door of Sam's office. He was very pleased that he did not have to engage in any discussion about why he is leaving that is remotely negative about the company or Sam. He felt sure that Sam's behavior meant he has not burned any bridges.

It was almost five o'clock and Greg ran to the Lab to see if Robert was still there. Greg wanted him to be the first one, after the meeting with Sam, to know about his decision to leave and wanted to discuss the transition efforts with him. He wanted to break the news sensitively to Robert before he hears it from others. Greg wanted to let him know that he is excited about the new opportunity, but that he will miss him and that he intends to do everything possible to make the final two weeks productive. Considering the delicate personal relation with him on the DDC3 test, being diplomatic may determine the perception Robert has of him.

Robert was not in the Lab and Greg decided to call it a day and drive home. And what a day it has been – the resignation from SONARES and the commitment to the new job at MICGEN has not quite sunk in. He ate a TV dinner and tried to distract himself from the events of the day, but he could not help agonizing late into the night, checking his e-mails several times to make sure that he would not miss any surprise notices that may be coming in from Sam or Vernon. The next morning he lay in bed, trying to stay as long as possible in that delicious zone of half asleep and half awake. And the memories of his day before flooded in his thoughts. Fully awake, he swung his feet to the floor and asked himself, 'is this all real'. He can hardly believe the events of the past two days. And then his mind shifted again to the troubled DDC3 system and he said to himself. 'I will help during your notice period, but my loyalty has to shift to my new employer and to their interests'.

Monday morning Greg went straight to the Lab. Robert was already sitting there chipping away at the to-do list that Greg prepared for him on Friday afternoon. He looked up and said "Wow Greg, did you work day and night on function tests last Friday to come up with this list? By the way, I had to take off most of the day on a personal matter. I was trying to find you to let you know and left a message at the reception, did you receive it?"

"Yes, I did" Greg said and feeling as if a knot was developing in his stomach, he continued, "Robert, I don't know how to tell you this, but I decided to take another job."

Robert looked up in disbelief and said "No, this can't be true, are you saying that you plan to resign from SONARES? I cannot believe that."

"I already did resign," Greg said.

"Is this job the reason for your move?" Robert asked.

"Actually, no. I was offered a career opportunity that simply was too good to pass up" Greg responded and hoped that Robert would not dwell on the problems with this verification task assignment, and he did not. He said "Well Greg, these have been challenging months for you and me, I wish you the best for your future endeavors." "Thank you Robert," Greg replied. They shook hands and Greg left the Lab.

Then, Greg made the effort to give notice to everyone else affected by his departure. He wanted to make sure that he personally told other key employees with whom he worked that he had resigned. And, Greg felt that he should say it in a way that 'thanked' the person for helping him develop his career. He made his rounds and said essentially to everyone, "I don't know if you have heard, but I am resigning to take a position at another company. Before I leave I wanted to be sure to let you know how much I have enjoyed working with you." Greg said to himself 'these people may leave for other jobs in the future and I want them to have positive memories of me. Who knows when they can impact my next career move.'

Later in the morning he received a call from the HR department wanting to know when it would be a convenient date and time this week for an exit interview. Obviously, Sam Watering must have informed them about his departure. Greg reminded himself to review all the documents he signed when he took the job at SONARES Engineering over six years ago.

He was reasonably certain that he did not agree to non-compete or non-solicitation clauses. Anyway, in this case it wouldn't have jeopardized his future. Whatever their reaction, Greg took confidence in knowing that the meeting with Sam Watering went fine and that he is well prepared both emotionally and professionally. During the HR meeting he essentially repeated his complimentary comments he made during the gathering with Sam. Then, he asked for them to review his performance with Sam Watering and to please give him more than the basic type (dates worked, job title etc.) reference. Although they said at the beginning of the exit interview that they would provide him with a reference, he wanted to make sure that he got a good written reference before he leaves.

Greg did not slack off. He felt that after giving notice of his impending departure, it was important for him to keep working hard and not coast for the remaining days. He told Robert "You can count on me to continue to do my job until I exit the company. Actually, since I know the situation with DDC3, I will step up my software verification effort." Furthermore, Greg left notes on exactly what the control system problems entailed and how he suggested handling the corrections. On the second project, a preliminary field instrument specification task, he documented what he did in detail so the information could be passed along to the person taking his place. Greg felt that doing so would show professionalism and that he still values the company.

Two Weeks Later

The two weeks went by fast and it was time to say goodbye to SONARES Engineering. On his last day, Greg left his office clean and neat. This way his replacement didn't have to clean up anything. Greg had a wrap up session with Sam, his boss, to go over the remaining duties and details on the two projects. He let him know that he is available for questions and gave him his cell phone number. Most importantly, he said thanks to him in-person a final time. He also wrote an e-mail to all the people that he has worked with, letting them know again that he appreciated working with them.

When he arrived home that day, Greg called Linda to tell her about his last day at SONARES, and as expected, he received some profound comments. After telling her that he was relieved that his last day at the engineering company went so well, and even though he is excited and looks forward to start at MICGEN, he is somewhat uncomfortable.

Linda said "Greg, personal growth is challenging. It is moving because taking risks is uncomfortable – the fear of the unknown and the possibility of disappointment stay in the back of your mind until you have adapted to the new company environment. The thing is, we manufacture this stress ourselves. Many of us fail to realize our full potential because we are too afraid of the future. I think you made a good decision. The path of success in a small company can be hard, but it also can be very rewarding; I know that from my own experience."

Greg planned to relax on Saturday to recharge his 'batteries' for the new job. Excited about the new job but also worried about things that may go wrong, he spent all day reading up on startup ventures. After all, Vernon just started the MICGEN Company only a little more than a year ago. So he was concerned about how he would be doing things in the new company environment. After scanning through several stories on Google, he felt more comfortable. Of course, Greg knew that he was not alone in that venture and he was definitely not in unchartered waters.

Being heavily interested in small businesses and how they work, Greg read statements from some of the founders of well-known startups. They talked about one of the toughest areas of starting your own company, what mistakes they made and how they overcame them.

Chapter 2 – CHALLENGES OF SMALL FIRMS

Monday morning, while eating his breakfast, Greg glanced through the headlines of the startup founders' statement paper which was still on his table from Saturday. Then he went for his morning run. It was pleasantly cool and he felt relaxed even though he was facing his first day at MICGEN. He said to himself 'I have known Vernon Averoni for years and considering how much Vernon was pushing me to join him at his company, the future working relationship will most likely be a good one.' He finished his run with a short stroll. He took his bath and dressed professionally, even though he has seen Vernon several times in casual when they met for dinner. He thought by being new, you never know when you will be called out to meet a key client. Greg judged himself to be capable, independent, and strong. Will this new job test him at every level? He hoped so, since he felt that his full potential was not used during the past six years at SONARES Engineering. He was ready for the fresh challenge.

He left early because he was not sure what traffic to expect on his route to the new office. As he drove, he could feel his increasing nervousness. He knows that during those first few early days where you are meeting everyone -- and everyone is meeting you -- first impressions about him and his future potential can make a major impact on his future success with this new organization. He also knows that nothing works better in all situations than having and expressing a positive attitude. He is determined to let his enthusiasm for being part of the new team show.

Toward the end of the trip to the office, which took about 30 minutes, Greg visualizes his first work meeting with Vernon. He had enjoyed a strong private relationship with Vernon, so he thought to himself 'don't let the prospect of Vernon behaving 'presidential' cause undue worry or stress. View the transition from private to business as simply another professional challenge. Your ability to accept it, better yet, to make the most of it, will enable you to stand out'. His thoughts shifted to Vernon's past comments about sales experiments and his new sales and marketing manager, Jim Barryson. He seemed to be excited about Jim and since marketing usually occupies the key position in a company, Greg felt that his working relationship with Jim may determine his initial standing in MICGEN.

He happened to beat the traffic and arrived earlier than he had anticipated, just a little past 7:30. He decided to sit out the extra time in the parking lot in an area away from the business's front door. Four cars were already parked near the building entrance. He observed the outside of his new office building, a small building compared to the huge engineering company facility he used to see, and said to himself 'well, this will be my new vocation home' since he often spends more time in his office than at home. He is used to being one of the first ones to arrive at the office and usually the last person to leave in the evening.

As he entered the building the receptionist greets him with a friendly "Hi, you must be Greg! Welcome to MICGEN."

Vernon must have told her to expect him and greet him by name. "Yes, I am. Thank you for the nice welcome. It is good to walk into a strange office on a first day and be greeted so nicely. Thanks again! Where is Vernon's office?" "Straight ahead, the first one to your right," she said.

Peering into Vernon's office through the glass door, Greg saw Vernon sitting at his desk seemingly deeply into reviewing some papers. Vernon looked up and put a big smile on his face. He stood up, rushed to open his office door and opened his arms to hug Greg. "It is great to see you here at my office. I am delighted to welcome you as the new Engineering and R&D Manager of MICGEN. Together, let's make the best of your years of experience in process control applications and your passion for exploring new system concepts." Greg knew that the hug was 'real' since he had known Vernon for years but was a little surprised how open Vernon expresses empathy and sympathy at the office.

"I appreciate this friendly welcome and look forward to working for you, boss," he said.

"Don't call me boss," Vernon responded. "You and I have known one other for a long time."

"O.K. Vernon, thanks for the opportunity you offer me at your company. I will certainly do my best." Greg replied.

"I count on it" Vernon said with a grin. "Let me show you around here."

Then Vernon was quick to note that before MICGEN entertained these new office spaces, he made sure that he had a concrete idea of where the business was headed. "And you are an important part of it" he added, patting Greg on the shoulder.

Walking out of his office, he turned right and said, "This is Jim Barryson's office. He is on an overseas sales trip and will be back Friday." They went a few steps further and Vernon said, "Well, this is your place."

"Very nice; my home away from home looks good," Greg commented while looking at the newly furnished office. "Yeah, that's what it is for me."

"This means that you need to make it as comfortable as possible," Vernon said. Then Vernon spent about 10 minutes showing Greg around the cubicle office floor, the lab and assembly room. It all had the trappings of a typical tech company; open layouts, standings desks and a small coffee area.

Vernon and Greg have similar interests and lifestyles. We got to know Greg a bit in the previous Chapter, so now let us introduce Vernon: Vernon Averoni is the founder of MICGEN. He worked at SONARES Engineering for five years and then co-founded SARAP, a safety instrument company, three years ago. His role at that company was Vice President in charge of operation, including sales and market development. A year ago Vernon acquired MICGEN, a bankrupt company with a Safety System product, serving the safety control market.

As President, he leads all facets of the business and built the company up from seven people to thirty-five in a short period. MICGEN evolved essentially from a customer project that not only provided for re-startup financing but also served as a testing platform of MICGEN's basic products. Vernon is known to have the level of commitment that only someone that "walks the talk" can live up to. He is in the office early in the morning, leaves late in the evening and is known in the industry for his brilliant and innovative technical and marketing know-how.

After showing Greg around the facilities, Vernon called a staff meeting to announce Greg's joining of MICGEN.

"I couldn't be more thrilled to have someone with Greg's level of expertise join our company, please welcome our new Engineering and R&D Manager – Greg Winkler," said Vernon. "We recently celebrated our one-year anniversary, and with more Engineers joining us every month, MICGEN is poised for significant growth. Greg is here at the right time, lending the experience garnered throughout his career to help take MICGEN to the next level; again, welcome my friend."

"I am pleased to be joining MICGEN at this stage of your growth," said Greg. "You are a solid young company with unique products. I know that you have an excellent management team with the added strength of experienced engineers in the safety control market. It's a privilege to be a part of your growth and to support MICGEN's vision of personal inspiration and product innovation."

Greg reintroduced himself to people he recognized from the meeting as they walked by his office and as he walked by their cubicle. He made eye contact, smiled and extended his hand for a handshake.

He knew from past experience how important it is to have a good relationship with everybody in the company. So, he said "Hello," "Hi" or "Good Morning" to let the person know that he is new at the company.

When he heard the person's name, he repeated it to help him remember it. In particular to the Marketing Assistant, Monica Campbellinis, whom Vernon mentioned when passing by her office, he said "It's nice to meet you, Monica." And, since it looked like she had enough time, Greg politely asked a few questions about her title and marketing duties at the company.

He tried his best to work on presenting a favorable image to new colleagues, to be professional and importantly, be himself.

He unpacked his briefcase and started organizing his office. Since Greg is one of those well-organized people, it was relatively easy for him to do so. Having a good-sized office allowed him to basically divide it into three parts – his desk, his worktable and the extension table where he could meet with clients or coworkers. All he had to do was to move the extension table and the associated chairs. It was a modest office, but he was pleased with the arrangement, which already provided some coziness at the new workplace.

Just as he was leaning back in his chair, Vernon looked into Greg's office. "Great. Looks like you already feel comfortable here. If you don't have plans for lunch, let's have lunch together around 12:30. There is an Italian place around the corner; is this OK with you?"

"Sure," said Greg.

"Well, then I'll go ahead and reserve a table for us," Vernon responded. Greg remembered that during the time, years ago, when he and Vernon worked at SONARES Engineering, lunches were a regular occasion to discuss their ideas and debate business matters, while at the same time enjoying the food accompanied with wine. From the onset, a 'business lunch' was never confined to a mid-day snack.

Food evokes a feeling of comfort within us. When we are comfortable, we can focus on our goals in a relaxed way to the point where things assume a more personal stance. This was always the case between Vernon and Greg.

So, while Greg expected some personal chit-chat to begin with, he knew this was going to be about business. Vernon has his agenda and his goals. Greg knows this is what lunch is about. And, he liked that about Vernon, communicating his plans in a relaxed atmosphere. They both liked dialogue, but for this lunch Greg planned to listen more than to talk in order to focus on what Vernon had to say. It was probably going to be a brainstorming session. Both had always felt creative with good food in front of them.

Sharply at 12:30, Vernon peeked into Greg's office. "Ready for lunch, Greg?" he said. "Let's take my car. I often have lunch at that place. I think you will like the food," he continued. As he drove, Vernon continued the conversation with "remember, back in our old days at the engineering company, the lunch with vendors was almost essential. Specifications were reviewed, delivery schedules were made, all over lunchtime martinis. Is this still going on when you left them?" he asked Greg.

"No, for the most part, those types of midday meetings are long gone," Greg responded.

"That's unfortunate; because with today's communication technology overload, the face-to-face lunch is still an important way to discuss delicate matters in a personal atmosphere and is perhaps even more valuable today than it was five years ago. Well, times change fast and not always for the better," Vernon concluded as they entered the restaurant parking lot.

At the restaurant the owner greeted Vernon at the door. "Hello Mister Averoni. Welcome again." "Thanks," said Vernon and pointed at me. "This is Greg Winkler. We have similar food tastes, and I hope that you have a quiet table in the corner for us."

"Of course, I do. It is your usual table."

And as they set down at the table, Vernon continued, "I think that lunch with a colleague or a client can often be more productive than an office meeting. Getting out of the office and off the phone creates an environment more conducive to relaxing and candid conversation." His five years in the leading company position have obviously convinced Vernon that meetings over lunch are a way to get a lot accomplished. Greg had the feeling that he could expect to have many business lunches with Vernon.

They ate and had some small-talk about family and friends. Toward the end Greg commented, "The food is really good here."

"Yes, it always is and their deserts are the best. Let's go for one," said Vernon and waved the waiter to order a ricotta cheesecake for him. Greg selected a ricotta pie. Vernon went on to tell a few short stories of failure in his previous company. Certainly the stories were entertaining, and he had fascinating theories about what he had done wrong and what was done to create a more successful product. Then Vernon asked Greg "what do you think of our products?"

"I have only looked at your safety system and that appears to be solid," said Greg

"You got it right. We have basically one product, the TMR-based safety PLC, consisting of the TMR CPU & Comm. module and the Intelligent I/O modules. The operator station portion is a trade labeled third-party creation" responded Vernon and added, "Our safety system has been really successful but almost from the time of its introduction customers keep on telling us that we need to incorporate some of the surrounding controls so that they don't end up with too many different devices."

"Yes," Greg responded. "Process plants are looking for complete application packages; single source solutions that are pre-configured and therefore easy enough to use for the typical safety and control engineer, not just for the highly skilled specialist."

"Yeah, our clients have been emphasizing this more and more. That is what I want to talk with you about," said Vernon.

"A few months ago I asked our product development team to come up with a new controller that incorporates both safety and control functions and interfaces with our existing intelligent I/O – our next generation product. One challenge for us at MICGEN is that the expertise of our engineers and programmers is limited to safety systems. The other factor inhibiting a simple evolution of our existing system is the rapid technological evolution of the components and techniques used. We seem to be wedded to the progress of the computer industry since we can rarely, if ever, get ahead of it. So I am concerned that when a few years pass after developing a new system, the components of that system may be termed 'obsolete'. By obsolete, I mean that the use of the latest components, techniques, and software would yield a better and more cost-effective system. Things are just changing so fast nowadays. How do you see all this evolving Greg?" asked Vernon.

"You are so right about the rapid hardware technology change, but I think that there are still key reasons to develop one's own system and most importantly, it is the software configuration that usually makes the success or failure. I would summarize it as follows," said Greg.

"From what I know, you do a considerable amount of business in the turbomachinery, accelerator and around other high-speed process areas. This seems to be your niche, and taken into account the specific technical requirements, there are often valid reasons for resisting the transplant of general purpose control systems. The other important reason to develop one's own control system is the need to have an intimate knowledge of the control system developed, since normally the developers are given the task of supporting, expanding, and improving the system throughout its lifetime. This reason sometimes rules out commercial packages," said Greg and continued.

"Again, regarding unique requirements, one of the reasons you may want to avoid third party solutions in favor of an in-house design is the truly unique requirements of the process equipment in your main niche business. The market for safety-related automation solutions with timing precision down to milliseconds is special. In addition, there is the fact that there seems to be practically no hard specifications available at design start and that the system must be able to evolve to satisfy a host of future rotating equipment requirements," emphasized Greg. "One not so obvious reason for developing a control system 'in house', is at times the need to avoid anything 'not invented here'. I believe however that this psychological motive does not apply to the controller and its software in this case. It probably really needs to be invented here; perhaps it relates to the operator interface. However, I do not know enough about the supervisory interface of the product to make such comment," said Greg.

"The operator interface, using Windows, was developed for general control and SCADA systems by a European company. It seems to perform reasonably well with our safety applications. This system employs a standard PC running a Windows GUI, again, it appears to be OK," said Vernon.

Innovative ideas for an automation system

Greg then commented, "I believe that the most significant aspect of any new development in our industry would be the invention of a process control wizard that combines multivariable control with constraint limiting and auto strategy configuration concepts, which would enable it to be used in an array of high-reliability type process applications. A few key processes, for example compression and combustion, could serve as a foundation upon which the application optimization work of other processes could follow. Pre-configured application software is not new, but the artificial intelligence required to enable the on-line adaptation to complex process units would really represent a breakthrough in our industry. The intelligence would have to reside in the controller in order to adjust to the fast dynamics during process upsets."

"Wow, considering the potential in petro-chemical plants where we already made inroads with our Safety System, this sounds like an incredible idea," said Vernon. "Do you think a small company like ours could develop such a complex wizard software module?"

"Perhaps," replied Greg. "Certainly, the large companies in our business could not, and most likely would not, do it. They are putting all the optimization software into their supervisory computers, which would not work in this case. Plus, they all have large application engineering departments, which sell assistance to their clients; thus, a development of this type would create a conflict of interest in their own firm."

"This would be a daring move for us. How can we assess the potential risks?" said Vernon.

Greg commented, "Yes indeed, since it has never been done before, it would be a challenging task to realistically evaluate such an undertaking. You mentioned that you requested the product development group to come up with a new controller. Have they, from a processor and memory standpoint, considered to include real-time optimization software?

Today processor and memory costs are no longer a competitive pricing issue. Thus, the principal challenge would be the software brains. We would not need a team for that, just the opposite. This would obscure it; only a single mastermind could convert this innovative theory into praxis. Vernon, I know of somebody whom his colleagues consider a real whiz kid, but let me go back to your basic question of risk valuation. While I have often given thought to process application auto-learning concepts, I need to study the specific applications further."

"OK, enough brainstorming for today. When we get back to the office, don't let me forget to give you that stack of information on the preliminary documentation of our new controller. It is only conception at this stage. So, when you comment on it, do not worry about impacting any cost or production schedules."

"OK, thanks for the lunch Vernon," Greg uttered.

"Hey, nothing to thank me for. Looks like we are in for exciting times. Let's have fun in pursuing our goal of the next generation process control and safety system," said Vernon.

It was already 3:30 p.m. and the waiter was preparing the other tables for dinner. Vernon paid and they left the restaurant.

On the way back to the office, during the five minute drive, they did not say a single word to one other. They both were trying to absorb their luncheon brainstorming. After all, they practically put the future of the company up for discussion. When they arrived at the office parking lot, Vernon said, "Greg, this was one of my most enjoyable luncheon meetings! What a great way to discover innovative ideas!"

"Thanks, Vernon. It was hard not to get inspired during our conversation," responded Greg.

As Vernon opened the door of his office, he said, "wait Greg, let me give you the preliminary notes on the new controller." Instead of a few pages of specifications, he came back with a foot-high stack of papers and said, "Please take a look at this, but don't bother today. Let's get together Friday morning to talk about it. Is this OK with you?"

"Sure," Greg responded, trying to hold the document stack together while he walked the short distance to his office to put it down on his desk.

He took a deep breath, looked at the stack of controller papers with a mixed feeling of over-challenge and happiness. He did not expect things to start at that accelerated pace. He was just going to listen to what Vernon had planned for him; instead he passionately put his idea of his 'super controller' on the table. Did he go too far? He was very pleased, though, that Vernon not only paid attention, but also seemingly understood him well. He felt comfortable in an environment where his boss had the 'finger on the pulse' and the necessary experience to make technical judgments, instead of SONARES Engineering where management was mostly interested in politics and schedules. Even though he has known Vernon for several years, this luncheon discussion re-emphasized Vernon's incredible entrepreneurial drive and Greg said to himself, 'I just hope that I can keep up with him.'

Greg was to some extent familiar with MICGEN's present safety PLC (programmable logic controller), a single large centralized system, so he was impatient to find out what the new system consisted of. He could not wait. Before he headed home, he had to pour over the controller documents to see the new architecture and to get a feel as to how far along they were on design. The papers found their way into Greg's briefcase for continued assessment at home. There they landed on the dining table.

On his last assignments at the SONARES Engineering, Greg had first-hand exposure to the technology evolution that led to recent developments in safety and control systems. Several suppliers had released new systems that significantly depart from those traditionally available.

Considering the powerful microprocessor and memory components, designing a new system might seem less complicated now. While Greg was not an electronics design engineer, his experience resulting from the extensive testing process of several of these safety and control systems helped him assess the basic do's and don'ts. That could be of significant value; especially since minor modifications at this early development stage could be accomplished without having much impact on cost and schedule.

Since he was not hungry anyway (the big lunch was still giving him a full-feeling), Greg set out to review the document package again. Bearing in mind that Vernon mentioned these were 'preliminary notes on the new controller', the documents appeared surprisingly comprehensive and detailed from a hardware perspective. Within less than two hours he was able to obtain an inclusive picture of the controller design. He typed his interpretation of the hardware design on a single sheet of paper so that he could use his analysis in the upcoming meeting.

Controller Hardware Architecture – Greg Winkler's Interpretation of Design Documents

The module utilizes a single-board architecture which includes the intelligence (2 microprocessor & memory), the communications (redundant Ethernets and a serial link), and the high-speed I/O serial link. Intelligent Termination Panels are used to connect the field - discrete I/O, 4-20 ma or intelligent transmitters/valve signals.

The design incorporates a module backplane which can accommodate up to three modules – allowing for non-redundant, dual redundant and triple redundancy. Such flexible redundancy scheme permits a safety and control system design that matches the reliability requirements for each loop in a cost-effective manner.

Note: The extensive safety experience of the hardware designers is obvious. The detailed information on the flexible redundancy is impressive.

Regarding the Intelligent Termination Panel (ITP), these seem to be existing units. Four of these ITP's are provided: one combination (with analog and discrete I/O) module, one discrete-only module, one thermocouple module and one communication-only module for intelligent transmitters. The ITP's are certified I-Safe. HART firmware is incorporated in the 4-20 interfaces. FF firmware is provided for the intelligent transmitter interfaces. Fieldbuses have been available for a number of years for process control. However, their use in safety systems has been questioned; but due to recent Fieldbus SIF product releases, that is changing. The redundant serial link to the controller is high speed (notes indicate megabit transmission with CRC and HSP security). An OPC Ethernet interface is also shown on the ITP, presumably for direct data acquisition.

Since each module includes its own redundant serial I/O, intelligence and communication, overall system size does not degrade the speed of response. This is one of the key points. It is certainly a promising design that would perhaps permit to put advanced control software down at the module level where it belongs. But is there sufficient memory capacity for the required data available?

The integral Ethernet communication (redundant Ethernet ports and communication processor on each module and a serial link as well) and the notes on the communication architecture indicate that an H2 Fieldbus may be considered. There was not enough information to confirm this. With regards to HART communication, there is nothing mentioned in the controller documentation; presumably, this is covered in the ITP. HART transmitters communicate diagnostic information over a standard 4-20 mA signal and have become quite popular.

While the documentation is not clear on the communication devices, there is almost nothing regarding the controller firmware in the package. The exception is a two page write-up on Sequence of Events (SOE) handling, which is assumed to be partially performed at the controller level. Required memory capacity?

The document package does not include any control function description, function listing, or explanation of the software configuration at all. There is no functional software design specification in the package. Some of the firmware concepts must have been taken into consideration, otherwise how could the development group have come up with their hardware design?

End of comments

Greg planned to ask the software team about this the next morning and then call an R&D meeting during the next few days. Certainly, before meeting with Vernon on Friday, he needed all the information to give his assessment on the design package some credibility. Satisfied with his document review progress, he decided to have a microwave dinner and then relax in front of the TV, where he almost fell asleep.

Early next morning at the office, on the way to the coffee corner, Greg bumped into Peter Moringsen, one of the software engineers on his team. He introduces himself again, and asked teasingly "well how long will it take to finish the software for the new system?" He followed up with "of course, I am only joking, I planned to get together later on this morning with you guys." Luckily, nobody was around because Peter appeared to be in a bad mood and Greg got an earful.

"We, as programmers, are constantly being asked 'how long will it take?'. And you know, the situation is almost always like this: The requirements are unclear. Nobody has done an in depth analysis of all the implications. The new feature will probably break some assumptions you made in your code and you start thinking immediately of all the things you might have to refactor. You have other things to do from past assignments and you will have to come up with an estimate that takes that other work into account. The 'done' definition is probably unclear. When will it be done? 'Done' as in just finished coding it, or 'done' as in 'the users are using it'? No matter how conscious you are of all these things, often your 'programmer's pride' makes you give/accept shorter times than you originally suppose it might take, especially when you feel the pressure of deadlines and management expectations. Many of these are organizational issues that are not simple and easy to solve, but in the end the reality is that you are being asked for an estimate and they expect you to give a reasonable answer. It's part of your job. You cannot simply say: I don't know." Peter barked.

Greg said, "Sorry Peter, I did not mean to upset you or put you on the spot here. I should have started off by expressing my appreciation that you are here so early in the morning, since it is not even seven o'clock."

"Oh, I have been here for over two hours and still have not been able to find the bug in that sub-program. I apologize for my reaction. What can I do for you sir?"

"Apology accepted. Don't let me hold you up much longer. But who is the main programmer working on our new controller?" asked Greg.

"Dave Drobe is our head programmer. He resides in California and works out of his home. He may have started on it but as far as I know, he is still trying to complete an alarm management software module on the existing product."

"Thanks Peter, and good luck on your debugging effort" Greg said.

Greg thought about how to best prepare for a new system project meeting because he probably does not want the software engineers' participation, considering that they seemingly have not started with the new controller task. He will wait to call Dave Drobe, since he is on California time, but in view of Peter's comments, he does not expect a positive response on the new software status. He knows that getting ready for the gathering is only half the challenge. He must also establish an atmosphere of leadership and communication and with the software situation he can expect a real test. He anticipated something like that already when he did not find any software planning documentation in the package Vernon gave him. Despite that, a kickoff meeting for the new system development project may be his best opportunity to energize the group and establish a common purpose toward progressing in the work. From his experience, he found that a worthy kickoff is the result of good planning.

The review of the hardware drawing package gave him a decent beginning, but he somehow needed to present the software aspects as reasonably normal, even though the indications are that they are not. Greg has a set of tactics that he uses to set a positive tone for the meeting even under controversial settings. After all, project start delays have not been uncommon at the place he worked before. The diplomacies help him stay organized, establish his leadership, and begin molding the individual engineer and designer participants into a team.

As far as getting a handle on the software schedule, almost every software engineer Greg ever knew chronically underestimated how long it would take to complete a task or series of tasks. Only the very best are able to give and meet accurate time estimates, while the rest are sometimes off by a factor of two or more. The problem is that, as generally creative people, software engineers often fail to anticipate the problems they will encounter. Even though many engineers will complain that product managers change their minds, almost none will account for that in their time estimates. No time is put in for meetings to go over requirements and make changes. Bugs? 'Our code is perfect and never has bugs, so we do not need to worry about that. After all, QA will catch anything we somehow missed', many believe. Some of the other engineers whom they rely on will be out? 'That's okay, someone else will pick up the slack', they often assume. All of these things add up to missed deadlines very quickly, but none do as much harm as the two main reasons things don't get completed on time: the lack of a comprehensive Functional Design Description and not factoring in time for learning. This goes directly back to a common flaw of programmers. They think they already know how to complete the tasks they are given, yet very frequently there are things they have never done before. The time estimates reflect a state of perfect knowledge, where you simply require a principal outline and plow forward. In reality, many tasks are more complex and are therefore frequently misjudged.

With this in mind, Greg called Dave Drobe. "Hello Dave, this is Greg. I just joined MICGEN as manager of R&D. Could you spare a minute?"

"Hi Greg, Vernon informed me about it. What can I do for you?"

"I understand that you are in control of the software for the new controller product. Would you mind sending me information about it, perhaps by e-mail?"

"Well there is not much to send. We are simply adding control functions to the existing product function library. I have started working on it but had to stop last Friday because of a commitment on an alarm management software module. This is a major task and I am not sure when I will be able to get back onto the new controller assignment," Dave answered.

"Could you forward the Functional Design Description of the new controller software by e-mail to me?" Greg inquired.

"There is no such Description and since we are only adding control functions to the existing software, I don't think we need one," said Dave with his voice getting somewhat irritated.

"Sorry to have bothered you. I recognize that you are under pressure working on several tasks. Would you mind if I call you back early next week?" said Greg.

"Sure, by then I should be out from under the woods with that alarm management program," replied Dave.

The conversation with Dave did not go as anticipated. Even though, after Peter's remarks, Greg should not have been surprised. He had, however, a hard time accepting that their lead programmer did not realize that the software of their present system cannot be simply applied to process control by adding a few control functions. Or perhaps Dave did not have time to analyze the applications and felt that the functions are needed no matter what the linking architecture is going to be in the new controller. Either way, Greg realizes that he faces a major challenge to convince Dave, and perhaps even Vernon, of the absolute need to have a detailed Functional Design Description for the software. Considering that the hardware design seemed to be quite far along from an architectural perspective, it was almost unbelievable that the software tasks did not appear to have been analyzed.

Greg was disappointed, so he went to Vernon's office. The door was open and Vernon was sitting at his desk with his head sunk into some papers. "Sorry to interrupt, Vernon, can you spare a minute?" Greg asked.

"Sure, what's up?" Vernon replied.

"Well I just talked with Dave Drobe and was informed that as far as the new controller software, we simply plan to add a few control functions to the existing controller and that is it. You know this will not work. Never mind about our brainstorming conversation on artificial intelligence, I am talking about basic process control."

"I am glad that you grasped this so quickly. I am sure that when you scanned through the documentation package I handed you yesterday, you were already wondering where the software part was. Dave may have looked into some functions only to familiarize himself with control. But you are so right. We cannot merely expand the software of our present system. The main reason for me wanting to have a meeting with you this Friday was exactly that subject. I did not want us to wait too long before discussing this."

"Yes, this is such a profound part of the new system. I am so relieved that we are on the same wavelength on this," Greg replied. "Thanks Vernon. And again, sorry for the interruption."

"Nothing to thank me for; looks like you are digging into the new system very fast," Vernon said and returned to reading his papers in front of him.

Greg went back to his office and followed up on his earlier decision to hold a new system kickoff meeting and handed the Meeting Agenda to each engineer of his group. He walked by their cubicle, made eye contact and simply said 'let's meet tomorrow at 10:00 a.m. at my office'.

Control System Development Kickoff Meeting Agenda

Date/Time: 10:00 a.m. Wednesday XX, X

Estimated Duration: One hour

Participants asked to attend: Everybody in Product Development Department

Purpose: Share the development status of the new controller product.

Goals and Deliverables: Discuss development tasks.

Project Plan: Introduction by Greg Winkler

Key Success Factors:

Communication Plans: Discuss plan to share information and updates within the group and interested parties.

Question and Answer Session:

Summary:

Note: Kickoff meetings set the precedent for the whole project. Kickoffs aren't just about creating ideas — they're also about setting expectations and minimizing surprises later on. There's also the group aspect to kickoff meetings. They are where you determine how you and your teammates will interact throughout the project, and they provide the opportunity to size up each other's strengths and weaknesses. But without the right collaboration, kickoff meetings are just expensive people discussing obvious things. In the next page, Greg's approach to a successful kickoff meeting for the new controller (in spite of the missing software portion) will be discussed.

The next day, everybody showed up on time at Greg's office. Greg welcomed all participants and explained that he intended to review the preliminary new product information and that they will have time for questions at the end. He then tells them that yesterday Vernon gave him a document package of the new controller and he has spent several hours reviewing it and believes that he understands the design architecture from a hardware perspective. He adds that he is aware there has not been a lot done from a software angle. He emphasized that if he misunderstood some basics to please interrupt his presentation.

He then handed out the note he composed last evening on the hardware scheme and said "This is my interpretation of the hardware design based on this documentation package," pointing to the paper stack on his desk. "Let me go through it item per item."

Greg glanced at the team and was satisfied to see that everybody's body language seems to express interest. He went through the design and feature assumptions and repeated some items to verify that his analysis of the documentation was correct. He complimented Paul Bingam, the lead hardware engineer, on the comprehensiveness of the documentation. He pointed out that he felt the memory size to handle SOE data and distributed intelligence might not be adequate and that there are potential bottlenecks in the communication portion. Also, he wanted to put off any discussion on the software tasks until such time as he had sufficient information. Greg did not want to get bogged down at this meeting, but felt that he should take the opportunity to relate where he sees problems can be anticipated.

Greg paused and looked at the group to make sure he still had their attention. He wanted to reinforce key success factors and explain why they are important, and elaborated by stating "Despite the fact that there is often opposition due to limited manpower, and I recognize that we are small group, it is my experience that a new product development process should include a detailed Functional Design Description of the product. Yes, for example in this case the hardware concept is laid out, but without the software description we cannot know if we overlooked something or not. We really need a comprehensive portrayal of the new product because there is such a big difference between our present centralized controller architecture and our new distributed design. This relates especially to the software area. The definition should also include verification and testing procedures." He continued, "And, the success of product development lays in the upfront detail definition virtually every time."

He then carried on with, "I also would like to remind everybody that teamwork is essential. We need to look out for one another. The objective is to complete the development successfully, and it is up to everyone to do their part and to help one another. Talk about assisting, I would like to make the following suggestion to share information and updates within our group: A bi-weekly development status meeting and the use of the company intranet to communicate anything which might affect the development progress and need to be documented. Also, for any major concerns do not hesitate to contact me directly."

Greg looked at his watch and said, "We have consumed less than 30 minutes so far, let me open up the meeting for question and answers."

"I have a question for Paul," said Donna, one of the programmers. "When do you think this new controller hardware will be ready?"

Paul replied, "I believe that we will have a prototype in about nine months. At this time we have barely finished the architectural design, and as you heard, there are still several basic points to be addressed."

Peter then asked Greg, "is Dave going to be the one doing the software?"

Greg answered, "Dave will need help. As you know, he is working on the alarm management task at this time."

Greg than summarized the meeting with a call for action on the Functional Description, telling them that he will speak with Dave Drobe next week to also discuss this with him, and thanking them for listening to him during this kickoff meeting. He then followed up to say that he is excited to be part of this development effort and looks forward to the first status meeting on the new controller in two weeks. He added, "I know that time is very valuable. We have completed this meeting in less than one hour. Thanks again."

As they walked out of his office, Greg said to himself 'I wander what they are thinking; this was supposed to be a kickoff meeting and I did not explain a basic project plan, neither did I make any statement about short-term and long-term goals that need to be achieved to realize certain deliverables, nor did I ask questions about the responsibilities of the individual participants'. Should I have told them that I purposely did not want to address these basic items because I felt there is presently a fundamental lack of understanding regarding the software of the new product requirements? Without the software information it will not be possible to define the hardware. Although almost nothing was accomplished from a technical perspective, the meeting was encouraging from a standpoint of getting across the importance of teamwork. Considering the drawing package, and the fact that the hardware and the software groups acted before as separate entities, this was probably worth the time.

Greg knows that he will need to go into application details in order to provide the team with a fundamental understanding of differences between the existing product and the new product. It is not a lack of intelligence or professional skills. Greg has a feeling that his team consists of highly capable people, but they live in their world of safety systems, which is, from a functionality standpoint, very different from continuous process control systems. He thinks of ways to best illustrate the functions of process control and comes to the conclusion that typical applications would make the clarification simple.

Greg planned to put together an Application Guide for Advanced Process Control after his completion of the initial product development tasks, perhaps starting in a few months, but it looked like his best bet in moving the software functionality requirements along was to immediately start with a few application illustrations of basic controls. Fortunately, this was his area of expertise and while there was no other reason to explain rudimentary controls than to convince his team of the software functionality necessities, at least it would not consume much of his time to put a short presentation together. In conjunction with a couple of flow diagrams from the Instrument Engineers Handbook, he felt that this approach would be convincing. He was determined to have that done by Friday. Thus, he closed his office door and started.

The next morning, as he was sitting at his desk still working on the application figures, Vernon came into his office and said, "Well Greg, I heard that you had a superb meeting with your group. Congratulations!"

"I did?" Greg replied and with a puzzled look asked "who told you that?"

"Both Peter and Paul said that everybody was impressed by your product analysis and by your teamwork emphasis."

Greg did not respond to the compliment, but thought to himself that 'they were probably happy because he did not put them on the spot with pointed questions'.

And then Vernon continued, "by the way, I had a phone conversation with Dave. The main subject was not about the new product, but he did let me know that he understood from the beginning that the configuration concept would have to be changed." Then Vernon added, "Dave is a brilliant programmer, one of the best, however, sometimes difficult to communicate with. But let's talk about this tomorrow." And Vernon walked out.

'It looks like I wasted my time putting these application leaflets together if the main software guy understands the concept already' Greg uttered to himself.

But in view of him being almost finished with it, he decided to make copies of what he had done so he can hand the information to Peter and Paul. He added a cover sheet with an introductory paragraph stating that he prepared this only to better illustrate the point he made during the meeting about the differences in safety and control function configurations. It may at least help increase the team's engagement and cross communication, he thought.

Greg went first to Paul's cubicle and said, "Hello Paul. I have put together a few illustrations with notes which show the configuration differences between safety and control loops. When you have time please look it over – no rush."

"Thanks Greg" says Paul. "I was thinking about your meeting comments and checked the memory capacity; we have space on the board to quadruple the size without significant cost impact" he added.

"That is great," replied Greg.

Then Greg went to Peter's cubicle and before he could mention anything, Peter said "Hello Greg. That was an outstanding meeting! What have you got there?"

"Well, thanks, Peter. I sketched out a few diagrams that may help clarify the control loop configuration. When you can spare a moment please take a look at them," said Greg.

"Hey Greg, since I have just been able to de-bug that subroutine, I will look at this right away," said Peter.

"Super!" said Greg and returned to his office. He was delighted. There are not only can-do attitudes evident but also signs that these key members of his team embrace change and will accommodate the new technology.

Greg arrived at the office very early on Friday; he was going over their present Safety System details in order to be prepared for comparison questions that may arise at the meeting with Vernon. Around 9:00 a.m. he went to Vernon's office to find out when the meeting was supposed to start.

"Hi Greg," said Vernon, his face lit up like a sunburst "ready for the meeting. Before we start I want to let you know I just received the good news from Jim Berryson that the big Safety System contract award by CAISTOS Onshore has been in our favor. Jim will be in this afternoon and I want the three of us to sit down and discuss this upcoming project." And Vernon added, "Jim brought back a request for bid to add process controls to one of our existing Safety Systems. This could be a huge opportunity to get our new system installed. We know this client well and can be open with them regarding our status; but let's discuss all that when Jim is here."

And Vernon continued, "Oh yes, and let me also talk to you about Dave Drobe before we start addressing other things, since this is such a key aspect of our new system. As I mentioned to you yesterday, I had a conversation with Dave. His estimate to get the Alarm Management done is six to eight months. I am putting a lot of pressure on him and must really apologize for him concerning the response he gave you."

"I guess this means that at this time we lack a lead software engineer for the new system?" Greg said.

"Yes, unfortunately this is the case; you talked about a 'software whiz kid' on Monday during our brainstorming session; is there a chance to get him to join us? What is his present situation?" asked Vernon.

Greg responded, "I must admit that I thought about this yesterday and I called Ken Beamer to see how he is doing, just a courtesy call; unfortunately, he said that he has an interesting new assignment. Ken is still young and fairly inexperienced, but you talk about creativity and dedication. Unlike Dave, which has many years of experience, Ken would need a detailed functional design layout upfront."

"Well, we probably need that anyway," Vernon interrupted.

"True," Greg said. "But it would be preferable for the software team to come up with that document themselves so that they are the stakeholders."

"Yes, I know, but in our case I am not sure how much those stakeholders really understand about process control requirements," Vernon said. "Why don't you think about that again and call Ken to see if there is a possibility of him joining us."

"OK will do," Greg responded.

"All right, so with regard to our meeting today," Vernon said, "Let me start with the product meeting agenda I prepared. I want you to have the background information as to what I had in mind with the new product; and if you have questions, don't hesitate to interrupt me."

Initial product planning

"To begin with, the product planning process was controversial within our small company because most of the team wanted simply to enhance the Safety System. Feedback from our customers and moves by the competition clearly indicated to me that this was not the way to go. We need to satisfy the customer requirements of the 'total solution' approach. With this strong client reaction, I decided two months ago to initiate a preliminary product planning effort, basically asking Paul Bingam to look at the latest DCS designs and to come up with a controller that combines the distributed architecture of a DCS with the reliability of a Safety System. The drawing package I gave you is the result of this. We have not done much more."

"Paul did an excellent job," Greg interrupted.

"Yes, I think he did," replied Vernon and continued. "I feel comfortable with our analysis from our customer's perspective, but on the sales channels I am somewhat unsure, since our sales strength is not in the application area. Regarding whether we have factored the trend in the control industry into our product plans, I would like to discuss this with you later - I mean after my short presentation regarding the following issues."

- "We have not really evaluated **discontinuation of our current product**. This is going to be a tough decision – cost of resources, obsolescence, etc. to maintain the product.
- Regarding **cost**, I have only a guestimate - we need to refine this once we have a handle on the software issue and after we have a functional requirements document.
- We also need **preliminary documentation** – A Product Description, so we can describe our new product to customers in terms of the value to the buyer, why it is better than the competition, etc.
- We need a **resource projection** – I worked with Paul to get an initial rough estimate on what it will take to test and build the product. But I met some resistance with this. It is understandable Paul did not want to sign-up to a schedule for a product of which software is not defined. So we really need to produce a Functional Specification that defines the product and an Application Guide that allows us to assess the market needs.
- And now, since we are going to have an opportunity to bid a project with a high probability to getting it, we need to

find a way to get to market in a shorter time than we anticipated."

"Greg, I realize you had less than a week to get a feel for what is going on here, what are your projections?" asked Vernon.

Greg replied "If you are asking for a prognosis, there is sufficient information from a hardware perspective to make a guestimate forecast, but as you know there is practically no software info and no lead programmer; therefore, my recommendation is to define the product from a functionality requirement, put together a sales bulletin and application guide so that potential customers and our sales channels know what the product is supposed to do, and to move as rapidly as possible to hire a software engineer. The most important action being the hiring of a programmer," Greg added.

Vernon then said, "well, then please make an attempt to reach Ken as soon as possible and if you are successful call me at home or better, here is my new cell phone number in case I am out of the house."

While it's only natural to want to celebrate the good news of the large Safety System order, Vernon knows that a big contract doesn't mean dollars in MICGEN's bank account--at least not for six months. It will probably require him to lay out more money for people and materials. He will ask production and test to put in extra hours to help the company get over the hump. Suppliers will also have to be contacted, etc. He will ask John, the project manager, to map out a project execution strategy, make a to-do-list, crunch the numbers and marshal the human and production resources. Of course the positive side of this large order is that he can pay down debt and re-invest some of the profits to expedite new product development and to grow the business.

At 3:00 pm Vernon called Greg. "Greg, Jim Berryson is here. Can you please join us in the conference room?"

"Will be right there," replied Greg and hurried to the meeting room where he found Vernon and Jim in front of the chalkboard. Vernon was putting up some diagram. Jim and Greg introduced themselves to each other and Vernon said, "Greg, I mentioned to you that Jim brought back a request for quotation to add controls to one of our existing Safety Systems. This was actually a misunderstanding. It is a Safety and Control system revamp job for AROBCO's Esmix, a complete offshore production platform in the North-Sea, a huge project. They are presently utilizing our ESD for the safety system and a DCS for the control system. They also have a supervisory computer. AROBCO likes our Safety System structure and they want to go triple-modular-redundant (TMR) on both the Safety and the Control system. This would eliminate the large DCS suppliers."

"Yes," Jim interrupted. "We have a real opportunity here."

And Vernon continued, "AROBCO is adding a compression unit. Thus from a time perspective we are talking about two years."

Jim then added, "they expressed to me that their present control system never worked satisfactorily from both a reliability and control standpoint and that the supervisory computer was down most of the time. They emphasized that they want to have a common TMR architecture."

Vernon said, "I have never been involved in production platform controls; are they complex Greg?"

Greg replied, "yes, due to cavity pressure changes and the injection systems, certain production platforms are difficult to control. They sometimes rely on their safety relief valves which is not only dangerous, but also gets the attention of the governments' environmental department. I worked on the control system design of one of those production units and must say that this was the most complex application I was ever engaged in."

"You have actually worked on the design of one of these units?" Jim asked.

"Yes, about four years ago SONARES Engineering did the instrumentation and controls for the CAISTOS platform," Greg answered. "This was one of the most challenging jobs they ever executed. I will never forget the control hurdles we faced. This is a long story; I can give you a presentation, which may be helpful for you in putting our proposal together."

"Wow," Jim shouted "that's awesome, how about Wednesday next week?"

"I want to be in the presentation. And Greg, can you get with me in about 15 minutes?" asked Vernon.

"Sure, I will be in your office then," replied Greg and they left the room.

Later that day, when Greg stepped into Vernon's office, Vernon stood up and closed the door behind him. "I need to tell you a little bit about Jim; he is a great guy and has a positive attitude. Regardless of the challenges he is facing, he always remains enthusiastic. If anybody is ever down, one can look to him for inspiration, and when the sales manager is a source of positive energy, that has an effect on the customers. Having said all that, unfortunately, Jim is not technically oriented, which is not a big problem in selling Safety Systems, but will be an issue with control solution selling. Which means that on a project proposal such as the one we discussed, you and I, would have to do all the work. Fortunately, we have six weeks' time to make this offer. You and I need to decide in a few weeks whether or not we want to make a proposition on this job."

"OK," Greg replied, "let's first focus on getting a programmer on board."

"Yes Greg," Vernon said. "Have a good and successful weekend."

Hiring of a programmer

During the drive home on Friday evening, Greg could not think of anything else but how he could get Ken Beamer to join MICGEN. He said to himself, 'I am far from an expert with personal interviews, but with Ken I have already ascertained that he is passionate about what he is doing, he communicates effectively in a small group, he has a good handle on the control function area, and I believe that everybody on my team would enjoy working with him. Yes, he has only a couple of years of experience under his belt, but his performance was outstanding on the job. But I know that even in the best of circumstances, hiring someone is hard. People are, as they say, complicated, and programmers especially so. Also, in this case much more important than what he knows is, how he learns it, and how quickly. And based on my experience with his performance, Ken has a good track record of learning new skills and applying them successfully.'

Saturday morning Greg called Ken Beamer. After three attempts he got through. "Hi Ken. This is Greg."

"Hi Greg, how is everything at your new company?" asked Ken.

Greg replied," great, that's what I want to talk with you about. I'll get right to the point. Since you have mastered that job your company sold to SONARES so well, I want to make you an attractive offer. The development you would be working on is about a new advanced control system. It would mean a fantastic career path for you! Ready for a super career move?"

"Wow, you got my attention. I certainly liked working for you. But what would I do with my piano?" asked Ken.

Greg responded "Well, our offer would of course include all moving expenses, including your piano. I am certain that we can make the situation very attractive for you. Also, here you would have increasing responsibilities, associated with it a higher salary. And, it's a new controller development; you would be getting in to it from the very beginning. It is a unique opportunity for you, Ken. By the way do you mind telling me what you presently make?"

"My annual salary is $79K, which includes benefits," Ken responded.

"That's a great salary, but I think we can do better than that," said Greg.

"I am interested!" Said Ken "Could you e-mail me a short description of what advanced control system you intend to develop?"

"Yes, I sure can. When could I get back with you?" asked Greg.

"How about Tuesday or Wednesday? Changing jobs would be a traumatic professional experience in my life, especially since they treat me well here," said Ken.

"I understand," Greg said. "I do believe that you can meet your individual performance goals much better at this small company here than at your present firm. The company is financially stable and is growing fast"; and Greg added "you have a good weekend Ken and I look forward to continuing this conversation next Tuesday or Wednesday."

"You have a good weekend, too," replied Ken and he hung up.

This new project potential also brought to Greg's attention that an updated company profile and preliminary marketing literature are urgently needed; whether they are talking with customers, putting together a proposal, or having a conversation with a prospective employee. It is essential to have this material. He notes that he has some marketing information from the competition. It could be used as a format. It needs to be solution oriented, and of course, it will be important to differentiate the MICGEN literature from that of the competition by stressing the basic advantages of the new system.

Greg plans to get with Jim Berryson and Monica Campbellinis, the Marketing Assistant, after his application presentation on the production platform system. This would most likely be a good time to make evident the urgency of requiring solution-oriented literature. In the meantime he will need to put something convincing together for Ken. He thinks that he can do that in a few hours tomorrow afternoon. Greg definitely does not want Ken to get concerned or suspicious because of him not being able to forward basic information about the product.

Sunday morning, after breakfast, Greg sat back and thought 'what should be in the brochure he will forward to Ken today?' Well, he felt, it really should be the same information as in the leaflet for the proposal package or the info which he would present to any potential customer. In other words, how can he make an effective brochure? How does he put together something that gets read and causes Ken or a customer respond positively? 'Every piece of literature sent out leaves an impression on our prospects,' Greg said to himself, 'whether it is a customer or a potential employee.' If he leaves the wrong impression with his bulletin, he runs the risk of losing Ken or in other instances, alienating a customer. Using the present MICGEN website content, which is strictly hardware and not solution-oriented, was not practical - except for the logo, the address and perhaps a portion of the head line and the service paragraph. Thus, Greg searched the main competitors' websites to examine their approach, while not having the intention to copy their content, but to see how to best differentiate his products and solutions. He knows that there are some basic literature considerations:

- **Understanding the customer** – He believes that he knows the niche market the new product is playing in very well. With that, he is aware of why they want to buy the product, what the crucial features are for their application and what the key problems are the new product can solve for them.

- **Attention** – How can he get the prospect, whether Ken or the customer, interested enough to read the brochure? Greg believes a sharp picture that relates to the customers' process may serve best.

- **Focus on Benefits** – 'Customers are not really that interested in our products,' he said to himself. 'They are interested in their business; thus, focus on the benefits they potentially will enjoy.' Although this does not apply to Ken's information needs, Greg believes that this needs to be the main message.

- **Headlines and Graphics** – He knows that the average reader takes just a few seconds to glance at the cover of a brochure and decides whether or not to read it. While this will not be an issue with Ken, a photo related to the process application and a headline that reads 'Advanced Control' is likely to get attention.

- **Benefit Headlines** - Use benefit headlines inside the brochure to hold their attention.
- **Bullet Points** - Use bullet points to focus on the key features – they will keep them attentive on what we have to offer.
- **Give them a reason to act now** – If he does not urge the reader to act now he/she may move on to the next thing that catches their attention. He will wait for Jim or Vernon to advise him what to best state on this.
- **Take away the risk** – Fortunately he has the outstanding reliability record of the existing system. He will highlight it, but surely the interesting party would want to know about the new system record – a little challenge.

He reviewed a number of brochure templates for designers on the internet and found that they are surprisingly professional. He felt that if Monica and Jim do not think so, they can improve the quality later, but right now he needs something that will at least do for Ken. Using the stock images he found so far may not cut it. After searching the internet for some time he found a sharp platform photo. He also found a good compression train picture he could superimpose. Then he shaded a snapshot of their existing Safety System and the whole thing looked pretty good to him. He also decided to use part of the prime MICGEN slogan – HIGH RELIABILITY changing it to HIGH RELIABILITY COMBINED WITH ADVANCED CONTROL – THE SOLUTION FOR YOUR PROCESS. While the leaflet content was otherwise very different, the format and layout resembled the existing brochures and the website, which may make Jim, Monica and perhaps also Vernon feel comfortable. After all, Vernon has built up a company which has a good product reputation.

It took a bit longer than he anticipated, but the e-mail with the attached brochure was on its way to Ken. And in the evening Greg had already received a response from Ken. "Looks like you already have the perfect system. Am I still needed (just joking)? – Brochure appearance is good! Talk with you Tuesday or Wednesday. Ken."

On Monday morning Vernon peeked into Greg's office "Hey, good morning. Any luck contacting your whiz kid programmer?"

"Good morning to you, too. Yes, I reached Ken Beamer. His first concern was moving his piano. No seriously, Ken is interested. He wants to take some time to think about it. I had to send him a marketing brochure on our new system. We need to phone him Tuesday or Wednesday. His present annual salary is $79K. I told him that I think that we can make an attractive offer."

"Well, great. That is a high salary though, assuming that we would need to add some percentage to it" Vernon said.

"Yes, he is making good money. I don't know what you pay our guys, but per statistics the median pay for programmers in the U.S. lies in the neighborhood of $76K per year. I think that we would have to come up with $89K and would say that Ken is definitely worth it," Greg said, adding "and let's not forget the moving expenses, the piano. Does our company have a relocation policy on moving furniture and other personal belongings?"

"No, does Ken have other household goods, furniture, etc. that need to be transported? Do we have to pay for re-tuning his piano?" Vernon asked half-jokingly.

"As far as I know, he lives in a furnished apartment so most likely everything he has will fit in his car. I will check with him to be certain," Greg responded.

"Do you think that we could negotiate on his salary?" Vernon asked.

"Well" Greg replied, getting a bit annoyed, "there are probably alternatives to hiring Ken, but the way I see it our software team is presently over their head involved with tasks on our existing product and I do not know any first-class software engineer, other than Ken, that we could hire to start on the new product. Do you have somebody else in mind?"

"No I don't. Let's go ahead with Ken," said Vernon.

"Hopefully he does not change his mind in the next couple of days. I will call him Tuesday and make the $89K offer to him. Is this OK?" asked Greg.

"Yes, go ahead," replied Vernon, "and all being well, maybe we can 'flange this up' this week."

"Did you say you sent Ken a marketing leaflet on the new system? What brochure did you use?"

"Since we do not have anything yet, I had no choice but to put a new brochure together on Saturday. It went only to Ken, don't worry," stated Greg.

"Can you please forward a copy of that new brochure to me?" said Vernon.

"Sure will do so immediately by e-mail," replied Greg and added, "I intended to wait until after Wednesday's application type presentation, which will likely highlight the urgency for sales and application literature, before showing the brochure to Jim or Monica. I don't want to leave the impression that I am taking over our Sales and Marketing tasks."

"I won't distribute the pamphlet," Vernon said and walked back to his office.

Greg gets back to working on his Application Guide for Advanced Process Control. He knows it will take weeks to complete such a document and is determined to use every 'spare minute' to make progress on this manual. In his opinion, this is not only the best way to illustrate the software functionality requirements but it can also serve as a key credibility document for a proposal, such as the bid that they will make for the safety and control system of the offshore production platform in the North-Sea.

And with that in mind, he paces over to Jim's office to ask if he could make a copy of the process diagram that was attached to the quotation request for that job.

Since Jim was not in his office, he peaked into Vernon's office and asked, "Vernon do you have a copy of the RFQ attachment, I mean that drawing showing the process?" Vernon looked up with a grin on his face and said "Hey, I did not know that you are an expert in producing marketing literature, this looks professional – very impressive also from a content perspective."

"Well, Monica or Jim may not think so, I am sure that they can improve the appearance, "said Greg and repeated, "I was trying to obtain a copy of that process diagram, Jim is not in; would you happen to have a duplicate of that drawing?"

"Yes I do. Here is the RFQ package. Just help yourself. Take out the documentation you need and have Monica make you copies," Vernon said.

As he looked over the RFQ drawing set, Greg realized that the project is almost identical to the job he worked on at SONARES. This was incredible! He was the lead control system engineer on that project and remembers the challenges well, including the difficulties they had with the supervisory control concept using a minicomputer. Also, he still recollects the pricing of the Safety System, Fire & Gas, DCS and TMC systems, separate systems which presented many integration difficulties resulting in cost and time overruns.

Greg was very excited. This was an unexpected break for him and allowed him to get back to working on his Application Guide this afternoon. Instead of having to analyze the RFQ documents and spending the whole day preparing for the presentation, he could write an introduction now and even include a system overview PPT illustration.

Proposal Introduction for
AROBCO's Esmix Safety and Process Control System

MICGEN is renowned as a leader in the field of high-integrity automation systems, and has designed and built several systems used in gas plants, refineries, petrochemical plants, offshore production facilities and similar projects.

MICGEN is quoting the MICWIZ system, a triple-modular-redundant (TMR) architecture. With it, we can provide a common hardware/engineering platform system for the Safety System (ESD), the Turbomachinery Control & Protection (TMC), the Fire and Gas system (F&G) and the Process Control System (PCS). The TMR based solution offers the highest availability and addresses such future issues as SIL/TUV certification.

MICGEN is able to assume responsibility for the entire AROBCO Esmix package - ESD, TMC, F&G and PCS, including equipment specification, project management, system design and manufacture, test, commissioning and training.

MICGEN personnel are experienced in designing advanced and reliable systems based around individual customer requirements. Our customer care capabilities provide comprehensive on-site services to ensure the smooth transfer of the system from the factory into operation.

MICGEN recognizes that application flexibility is the key to the success of this type of Production Platform. Our MICWIZ system incorporates a patented Advanced Control Wizard (ACW) which will minimize compression train interaction, assuring for smooth and safe operation under all operating conditions.

Overview of the MICWIZ Control System:
The proposed system will perform both the Emergency Shutdown and the Compressor and Process Controls. The TMC system will consist of the same hardware as the ESD, F&G and the PCS

systems. Result: A common high reliability system platform and architecture for the whole AROBCO Esmix production platform.

(Insert PPT overview illustration here)

Tuesday morning at the coffee corner, Greg sees Peter who instantly greeted him, "Hello Greg, I heard you are giving a presentation tomorrow morning. Do you mind if I sit in?"

"Of course not, I was going to invite you early tomorrow anyway," said Greg.

"Paul is also interested," said Peter. "I certainly will invite him as well," said Greg and stressed, "it is going to be about a potential quotation for a production platform system; hope it's not too boring for you guys. Thus far, Vernon has not told me at what time he would like me to start."

"Jim said the presentation will be at 10:00 a.m." replied Peter.

"Thanks, now I know the time as well," said Greg and carried on, "anyway what are you up to so early this morning Peter, another challenging debugging effort? I sure appreciate you being so dedicated."

Peter, complacently, replied "I have been here for a while; no, this time it's not about debugging, I am making good progress on that preauthorization software routine and just want to get it done so that I can start testing Monday."

"Great, don't let me hold you up," said Greg and thought to himself 'with these types of people we may pull off our new system development in time. It's not even 7:00 a.m. and he has already been here for a while.'

Greg returned to his Application Guide chore. He has difficulty concentrating, questioning if he should call Ken Beamer now or wait until the evening. Although he is starting to feel more comfortable with his local team, especially Peter, so much depends on getting an additional software engineer on board. Hiring the right people is critical for any business, and that's especially true for a small company like MICGEN with relatively few employees. Greg absolutely needs Ken, who he knows is first-class, to come to work for him. And he is concerned that he was trying too hard to sell Ken on MICGEN. Greg said to himself 'don't sell too hard on that upcoming phone call, even if you're desperate; trust that Ken will recognize the opportunity'. So he decided not to push further and wait until tonight or tomorrow morning to call Ken.

He did not feel like preparing dinner at home and went to eat out. He got back home late. At 9:50 p.m. he called Ken. "Hello Ken."

"Hello Greg. I have been waiting for your call," Ken replied.

"Great, hope you have not changed you mind," said Greg.

"No, the more I thought about it, the more I liked the idea of getting involved in the development of a new controller at the beginning stage. I went through your literature over and over and believe that you have a fantastic control module defined. I spent the weekend at my parents' farm and they are very concerned; but I told them that sooner or later I will have to move on. They seem to understand. So, on the basis of what we talked about on Saturday I am ready for the move, provided that you have a reasonable offer for me."

"I have an excellent deal," said Greg. "I have discussed the situation here with Vernon Averoni, our President, and can formally extend you an offer of $89K salary. Your position will be software engineer. You would be reporting to me and ah, we pay for moving your piano."

"Super, I take it. I am excited to join your team," affirms Ken.

"That is great! So if you are good with the offer and this is a verbal acceptance, we will put together the formal offer letter and get it out to you by DHL to your home address. A description of the benefit package will be included," said Greg, and he continued, "when would be your starting day?"

"I need to give at least two weeks' notice here and it may take a week longer to finish the program I am presently working on. I don't want to leave them in a bad situation," responded Ken.

"I understand, but please don't make it too long," replied Greg.

"Don't worry, you can count on me. I will finish the task here most likely within the two weeks," Ken said.

"Ah, before I forget, do you have other personal belongings, besides the piano, which we need to move?" asked Greg.

"No, I live in a furnished apartment here and I can put everything I have in my car."

"OK, we will look for a place for you here. Yeah and another thing, please sign and return the offer letter within a few days of receipt; actually, if you could send me an e-mail confirmation before mailing it, I would appreciate it," said Greg.

Ken also mentioned that he has his own PC and that Greg did not have to provide him one for him. Greg and Ken then chit-chatted another five minutes about their experiences on a job. Both where relaxed and felt that they accomplished a lot. The hiring of Ken was a major milestone in getting started with the software development of the new system. Greg informed Vernon next day and expressed that this would, in his opinion, make the bidding of the large production platform system more realistic and that he feels positive about that potential project.

Shortly before 10:00 a.m. Greg went with his computer, copies of the bid letter and the page of the literature thoughts in hand to the conference room. He had six PowerPoint slides prepared: one System Overview, one Process Flow Diagram and four MICWIZ brochure slides. Jim, Monica, Paul and Peter entered within a few minutes and Vernon also showed up shortly afterwards. Greg started by putting up the System Overview slide. Everybody leaned frontward, indicating a certain surprise. "Hi, everybody, this is the system we would probably be proposing for the AROBCO Esmix project; that is, if Vernon elects that we submit a quotation." He looked at Vernon, and then said, "Vernon said that he will make a decision in the next two weeks. And this is my suggestion for our proposal introduction letter." Greg handed each a copy of the letter he prepared and said, "while you read it I should tell you that I worked on an almost identical project three years ago. That is the reason I was able to produce a few slides and the short write-up."

Jim was the first one to look up after reading the letter, he had a smile on his face and said "Greg, I think we got that job in our pocket. Who at AROBCO would turn down an offer like this?" he added jokingly.

Vernon stepped in to save Greg and said, "Greg, this is outstanding, a detailed drawing showing our future system for that specific application and a very good bid letter."

"Wait," said Paul, still being focused on the slide, "what is the device connected to our controllers?"

"It's the Bently Nevada vibration monitoring system, a third party item," replied Greg.

"Yes" Peter said to Paul. "We have been interfacing with this package several times already, remember."

"Let me go through the System Overview with you and then we will review the next slide which presents the Process Flow Diagram," Greg said.

He started explaining, "the headline on the System Overview slide states:

"MICGEN Integrated System – Common TMR Platform for AROBCO Esmix – PCS, ESD, TMC, F&G"

Then Greg paused and looked at Jim. "This is what the client is asking for, isn't it Jim, an integrated solution on a TMR platform?"

"Yes. That's correct," Jim responded. Then Greg turned his head to Paul and said, "Paul, from your drawing package it is my understanding that this is what we plan for our new system, regardless whether we consider the AROBCO Esmix project or not."

"True, Greg," replied Paul.

Greg then described the application of the individual MICWIZ modules.

Following the description of the System Overview PowerPoint slide, Greg displayed the Process Flow Diagram of the AROBCO Esmix production platform. He said "I have seen this diagram so many times during my previous project assignment, which I mentioned to you, that I almost could walk you through it in my sleep." He then explained the process in general terms. He emphasized that the compression part plays a key role in this process and that he visualized MICGEN's future software development will include automatic configuration for turbomachinery applications. He did not want to spend more than five minutes describing the process because he knew that at this meeting everyone, with the exception of Vernon, would not derive much benefit from a detailed process clarification.

"One important item I need to add to my presentation" said Greg "Is literature. Whether we are assembling a bid such as the one for AROBCO Esmix, or we speak with customers, or attend a trade show, we will need literature for our new product. The printed word can carry credibility and help us communicate and strengthen our message, especially in our situation where no physical evidence exists on the new product." Greg, at this point put up his PowerPoint MICWIZ brochure slides and said, "with the increased availability of powerful desktop publishing systems we may be able to meet these needs internally, at least on a preliminary basis. I prepared this four-page brochure last Sunday."

He carried on, "but I have to add that I had difficulty finding quality photos on line and therefore the cover page does not, in my opinion, look as professional as it should. Of course in this case I did not have a picture of the product, so the background photo may not matter." Greg continued, "so perhaps for some literature pieces we should resist this impulse in doing it ourselves. I am certainly no expert in this. Our homegrown materials may betray their off-the-cuff origin to some of the people who read them. Appearance is reality in marketing and we have to look professional. But as an interim solution, until we have the real product and sharp photos of it, we may want to bridge the gap with our own creations."

Jim raised his hand and said, "Greg you really undersell yourself, that brochure looks good. I must point out to you that we do hardly any literature work here. We still utilize the literature from the previous owner of this company. Yes we changed the logo and the address, but that's about it."

"What do you do regarding the Website?" asked Greg

"We had a website development firm create it. It was not that expensive, and we now maintain it ourselves," Jim said. Monica added, "I took an online course and it is not that difficult to keep our site up to date. However, now with the new product we will need quality photos, etc., just what you have mentioned regarding literature."

"Thinking of what we need for the new product, literature is going to be a real challenge, especially given that we need several pieces almost now," said Vernon. He added, "some of it we probably have to do in-house just to meet our schedule requirements. Our present literature covers most technical aspects of our existing product and is totally hardware oriented; we even lack marketing emphasis there. Considering the scope of our literature task - updated company profile, new product data sheets, application guide, PowerPoint presentations, Website update, and later a selection guide, case studies, white papers etc., we need to discuss this separately. But right now I need to make a phone call," and "Greg and Jim let's meet in my office in about ten minutes."

Greg thanked everyone for their attention. He was going to hand out the page he typed on literature thoughts but considering how the meeting evolved, he felt it was better to keep this to himself. Everybody, except Vernon, went to the coffee corner where the conversations about AROBCO Esmix continued.

Paul said, "this was impressive Greg, what is the dead line for the submittal of the proposal?"

Greg replied, "it's really Jim who has the details; Jim what are the time schedules on this job?"

"Well, the bid is due in four weeks. The good thing about the project is that the delivery of goods would be about two years down the road," replied Jim.

Then Peter interjected with a sense of humor, "well, that will provide at least opportunities to deviate from what has to be done and explore crazy alternatives."

"Be serious, Peter," said Paul and added "in the timeframe of about four months we could come up with an MICWIZ prototype."

"Oh really?" said Greg. "You mean with slightly modified existing software?"

"Yes, Greg," replied Paul.

"Let's get together on this - you, Peter and I, tomorrow morning," said Greg.

Paul and Peter started to discuss adaptation necessities of the existing product and Jim and Greg went to Vernon's office.

Vernon just put the phone down as they walked into his office. He ran his right hand back and forth at the edge of his table, apparently trying to find his words. "Well, we have a lot on our table – the big Safety System project is going to occupy almost everybody's time here, and frankly I don't know how we are going to wrestle such a task as the new literature in the short timeframe we have, and I recognize that these are real needs."

"Yes and right now the needed literature serves both as definition for our new product and as credibility for our proposals involving the new product," said Greg.

"Yeah, I know. It's an important determinant of the product development success," replied Vernon.

Greg added, "there is no other option but to focus for the next few weeks on literature; I can do some of that after hours, but with regard to using this literature for marketing and sales purpose, we may have quality appearance issues. Would Monica be able to help me with photos and illustrations?"

"Yes" said Jim. "I was going to suggest this; Monica would be able to assist, and while she may not understand the technical content issues, she is very good when it comes to improving appearance and layout." Jim added, "Since I will be traveling overseas again starting tomorrow, I cannot contribute anything on this subject. By the way, do you mind if I use your brochure Greg; I can have it printed at FedEx this afternoon and take it with me."

"Go ahead, totally up to you Jim," Greg replied.

Vernon concluded "OK guys, maybe we can pull all that off in spite of our present overload, thanks!"

Over the Next Few Weeks

For the next few weeks, the workplace environment at MICGEN was what one could define as 'controlled chaos!'- having too much to do with too little time to do it. This problem was mainly caused by the CAISTOS safety system project award, a job that was almost twice the size as MICGEN's annual sales figure. Things such as unrealistic deadlines and increasingly heightened expectations were common causes of disordered multitasking, uncertainty, and interruptions during work. While it did not affect Greg much, because the many after-work hours he spent on literature and configurations were tasks he defined himself, several other people had problems with the long hours of chaos in the workplace, including the project manager, John Kramo. The difficulties encountered when trying to juggle the demands of clients and superiors with the needs of subordinates has the potential to provoke a lot of stress, and in his case, John ended up in the hospital. Fortunately, Vernon could convince a retired friend to fill in temporarily and the project and production schedules were not impacted.

During the past two weeks, from the time of his employment acceptance until his start at MICGEN, Ken Beamer had several evening phone conversations with Greg. He was already mapping out the basic MICWIZ controller configuration scheme based on Greg's definition details. Although Greg kept emphasizing that his description was preliminary, Ken could see his way through the software arrangement. Greg was delighted and when the time came to welcome Ken at the office, he ensured that Ken felt at ease, comfortable and supported. The welcoming processes included an office tour, introductions to the development team, the double desk Ken wanted, and an explanation of apartment options by Vernon's secretary, - all on the first day. In the first couple of days Ken also got information about the existing products and services, customer service philosophy and the big CAISTOS project initiatives, with which almost everybody was occupied.

Chapter 3 – THE PROCESS CONTROL WIZARD

While he was working away on the Application Guide, defining the control functions and how they relate to the different processes, Greg could not get his mind off the challenges he experienced on the controls of the CAISTOS production platform, which was almost identical to AROBCO Esmix. He had been involved in the automation solutions at all levels of that process, gaining in-depth experience. He had worked side-by-side with the customers' engineers and operators. He was able to implement some advanced process control and it was overall a successful project. But he was frustrated because he was unable to apply the multivariable model-predictive control the control system supplier specifically provided in accordance to his definition. He was convinced that this would have not only facilitated the operators' challenges, but would have also resulted in fewer process upsets. Due to the large company environment and the politics on that job, he was powerless to apply his ideas to the full extent.

The more he thought about this lost opportunity, the more determined Greg became to pursue the AROBCO Esmix project. To convince Vernon to bid the job was the first step. At this time, with the big Safety System order in house and the company already being overextended, Greg knew that he would have to come up with hints that their proposal effort would not be an exercise in futility. With no product to show and no installed reference he could only emphasize his process know-how and his outstanding working relationship with the client of the CAISTOS Production Platform. Greg decided to refresh his knowledge by analyzing the process again. He would call Hank Sandover, the Operation Supervisor of CAISTOS, whom he knew personally very well, to find out how the controls were performing and to see if he knew anything about the upgrade of AROBCO Esmix.

Greg took the process flow diagrams home. He did not have to spend much time on the control methods; he still remembered the implementation of most loops. Early next morning, because of the seven hour time difference, he called Hank, who picked up the phone on the first ring. "Hello, Sandover here," he said. "Hank, this is Greg Winkler, how is everything with you?".

"Great to hear from you. I see a 001 code. Are you calling from America? What can I do for you?"

"Yes, I am calling from Houston. I was wondering how the separator pressure control is performing and if you had the stage control valves installed? I am looking into the controls of the AROBCO Esmix platform."

"Really - I understand that they are adding a compression stage and are redoing all the controls. Ray Villaloberg, your buddy, left us to join them; give him a call, he will be glad to hear from you. Here is his new number xxx xxxxxx. Yeah, we put in the stage valves and it works better. But, since you are not here anymore, we don't have anybody to listen to our recommendations."

"I will give Ray a call, thanks for the advice," said Greg. "You are welcome and take care," responded Hank.

Greg was thrilled; this was almost too good to be real, Ray potentially being his client as it was the case two years ago.

Greg called Ray right-away. "Hello Ray, this is Greg Winkler; I got your new work number from Hank."

"Well, hello Greg. What a surprise. How is the world treating you?" asked Ray.

"I am doing great Ray. Just recently changed jobs, I am no longer with SONARES. I am with MICGEN Controls now, and at this moment I am looking at the AROBCO Esmix Process Flow Diagram."

"This cannot be true, we may be working together again; I would really look forward to that. I understand that we have your Safety Systems; they seem to hold up well. I am new here, so I am not very familiar with things yet. Greg, I am late for a meeting. Can you please call me at home? Do you still have my number?" asked Ray.

"Yes I do; will talk later. Have a good meeting" replied Greg.

Greg went at once to Vernon's office. Vernon had an expression on his face as if he were listening to something.

Greg said "sorry to disturb you Vernon, I just wanted to let you know that I feel we may have a realistic chance on the AROBCO Esmix bid. I just spoke with a former colleague, he works now at AROBCO. I will talk with him later again. If they feel strong about TMR, would we submit a proposal?"

"Yeah, we should be putting an offer together, but considering our workload at this time, I don't know if we can. Could you do most of the proposal work?" replied Vernon.

"Yes, I don't think it will take me very long to get familiar with our bid format, and regarding the technical part, I know what to put in the offer. I would need a lot of input from you on pricing though" Greg said.

"OK, sounds good" said Vernon and his tiredness and worried look told Greg to get out of Vernon's office. The huge Safety System job had also taken a toll on Vernon.

Greg phoned Ray Villaloberg at lunchtime. It was 6:00 p.m. for Ray. He started, "hi again, Ray, hope your meeting went well."

Ray said, "hi Greg, yeah no problem. I still can't believe that you are working on our Esmix Production Platform revamp job."

"Well, we are not working on it yet, I am only looking at the RFQ documents. We are expanding our system to include controls and since we have a Safety System installation at Esmix, we intend to submit a proposal," replied Greg and followed up, "you probably look at us as a Safety System supplier. Do you think that your people would seriously consider our proposal for a complete solution - Safety System and Control System?"

"After my meeting, I did some asking around here; I still have to feel my way through here since I am new. Anyway, they have many problems with the present DCS and want to go with TMR architecture for both the new Safety System and the DCS. So it appears that you have an opening here," said Ray and carried on, "I don't know if it shows on the RFQ drawings, but we are adding a compressor train. So, you are talking of a plant start up with the new system in about two years, if everything goes well with the main equipment delivery."

"Well, that would give us plenty of time to implement and test our new process control functions," said Greg.

"Talking about controls," said Ray and asked, "can you spare a minute to chat about the compression and separation units?"

"Sure," said Greg, "go ahead."

"Well, we seem to have a more severe problem here with our separator interference than we had at CAISTOS. I have been asked to look into it. I will request that they add the stage valves because this helped at CAISTOS, but there are still dynamic process interference issues which under certain process conditions can cause the separator relief valves to go off, and in two instances they resulted in plant shutdowns," said Ray.

"I remember the situation at CAISTOS very well," responded Greg.

"This is what I would like to discuss with you again. It will probably take some time, so should we talk about it now or would you rather talk it over some other time?" said Ray.

"No this is as good as any other time for me," responded Greg. "I am listening."

Ray articulated, "Greg, I believe that your MPC building block, which essentially consisted of a multivariable process controller with hi-low constraints and fallback features, would have helped to eliminate the process interaction issues.

Recall, we were pushing to include it in the Cobos DCS, but project management did not approve the change order."

"Well, Ray, I too think that it would have improved the process fluctuation problem, but remember Cobos wanted to incorporate the Module into their supervisory level, which I am almost certain would not have worked due to speed of response limits. This function belongs down into the controller and their device did not have the memory capacity for that," said Greg.

Ray then responded, "Greg, I need to ask you a loaded question. With your application knowledge and MICGEN's hardware expertise, could you guys come up with something like the MPC in your new controller?"

"I am pretty confident we can," said Greg.

"Can you get back to me on this in a couple of weeks?" asked Ray.

"I certainly will," ensured Greg and they got off the phone. Greg said to himself 'this was one of the best phone conversations I ever had'.

Greg knew the first thing Ray would ask him when he called back, would be 'when can you deliver the Multivariable Process Controller (MPC)?' His feeling was that Ray would not find a nine-month delivery unreasonable. And, taking Paul's statement of four months on an Advanced Control Wizard prototype module into account, this may be within MICGEN's capabilities, provided he could come up with the application definition for the complete MICWIZ Advanced Control Wizard. Since the MPC is an integrated software piece of the Application Wizard, Ken would need to have the detailed scope of work in order to start implementing the Multivariable Process Control software. With the existing CAISTOS MPC functional description, Ken should be able to give him a programming estimate.

Greg remembers well that he has the Multivariable Process Controller functional description, which they planned for CAISTOS, including many software details, in his possession. Even though the MPC was mostly his definition, and he is reasonably confident he did not sign a confidentiality agreement, he is not sure that, from a legal perspective, he would be able to use this detailed documentation package. Not only has he spent many weekends working on MPC during his time at SONARES, the head programmer at Cobos completed most of the coding and Ray put in a lot of time working on the details with Cobos. This was a task that took several months to complete. And, the document provided not only the information of what to build, it also delivered to the test engineer the details as to what tests to run. Greg knows that he would need Ray's help in sorting out the legal aspects.

While Ray Villaloberg is cognizant of the existence of this MPC document, he most likely has a copy of it, and pushed hard to have the MPC incorporated in the Cobos DCS at CAISTOS. He, being very software cognizant and detail oriented, was not effective in presenting the features and benefits to his management and the approval to put the MPC into plant operation was denied, even though CAISTOS has already paid for the development effort. Greg remembered

117

that and realizes that he needs to provide an overview type write-up to enhance Ray's ability to convey to his colleagues at AROBCO what they would be getting. And since he intended to include the Advanced Control Wizard (ACW) in the quotation of the Esmix Production Platform revamp job, he contemplates to present the complete Application Wizard software module to Ray when he phones him in two weeks.

Greg puts together the following narrative –

A New Generation of Systems

Advanced Control Wizard – ACW

The Advanced Control Wizard (ACW) is a module that provides control for a wide variety of applications – from simple control to advanced control, to total process unit control optimization. Its architecture is revolutionary.

- Each ACW module is of single-board design; with processors, memory, communications and serial I/O links.
- The XMR redundancy architecture provides for: Single – Duplex - TMR – and Quad redundancy levels.
- The ACW module is connected to an I-Safe Intelligent Termination Panel (ITP) via redundant serial links.

In addition to the standard control functions, the ACW incorporates three software elements:
MPC - Multivariable Process Control, **CLC** - Constraint Limit Control, **PCG** - Process Configuration Genius

The Advanced Control Wizard (ACW) includes an integrated set of desktop tools, the Workbench Wizard, that guide through the process of building, testing, and deploying the controls and associated applications.

Multivariable Process Control - MPC

Multivariable control is not new. It has been applied since the 1990's. MPC has demonstrated the ability to uphold certain processes at their optimal operating point. Its success has been limited, because the MPC's on the market interface to processes via the DCS or by means of a process information management system. Therefore, the existing MPC's have been mostly applied to linear control problems with slow dynamics. Since the MICWIZ MPC resides in the distributed controller, it has direct process access with millisecond response capabilities, eliminating most dynamic limitations and reducing process inconsistencies across the entire operating range. Enhancements also incorporate a look-ahead algorithm and more flexibility/simplification for building the model. The result is good process disturbance rejection for both slow- and medium-dynamic processes.

The MICWIZ MPC firmware encompasses:
- The online control program which executes input validation and the look-ahead algorithm, as well as the steady state target calculations and the dynamic move calculations.
- The multiple model identification algorithms plus model prediction, model uncertainty and cross-correlation features for model analysis.
- The control configuration wizard module.
- The simulate program which enables interactive evaluation and testing of control performance in case of model mismatch and process measurement noise.

The achievement of existing MPC's has also been restricted because most of the process control systems (DCS, PCS, etc.) on which the MPC resides do not have the proper safeguard and automatic fallback strategies that provide for loop integrity and reliability in case of certain field measurement (transmitters, converters, analyzers, etc.) malfunctions.

Constraint Limit Control - CLC

With Constraint Limit Control one can safeguard advanced control solutions for a wide range of processes, from simple linear to complex non-linear. This is a leap forward in multi loop integrity and reliability.

The MICWIZ CLC firmware contains:
- The Histogram and Normalcy Probability Plots for Continued Normality Test
- The Measurement Discrimination Evaluation
- The What-If Analysis Routines
- The Constraint Soft and Hard Limit Value Calculations or Pre-Settings

Process Configuration Genius - PCG

The automatic configuration concept of the Process Configuration Genius provides a new level of control strategy flexibility and efficiency. It integrates the application pre-configuration tools with on line control strategy monitoring and automatic selection of fallback strategies. It enables optimum strategy adaptation through interactive evaluation of process unit performance.

The MICWIZ PCG firmware comprises:
- The Process Unit Efficiency Calculation Program
- The Process Strategy Library
- The Expert Tuning Parameter Calculator
- The Automatic Control Fall-Back Strategy Selection

The Advanced Control Wizard – ACW – is a breakthrough in control technology. It provides advanced monitoring and control for real process optimization in a safe and reliable manner.

When Greg arrived at home that evening, he took the two boxes stored in the garage, marked SONARES, and searched for the MPC document. It was on top, in a thick folder in the second box. He flipped through the Functional Specifications and set them aside. He would review it tomorrow morning. Since Greg had two weeks to get back to Ray, he did not intend to interrupt Ken with the complex task of the Multivariable Control until such time that Ken had the basic function execution in place.

When he went through the Functional Specifications of the MPC, for which he did the basic definition, he realized again how much detail was necessary to provide adequate implementation and test information for the programmers and for the QA engineers. User interface details down to the pixel and color shade, size and allowable contents of data input fields, exact text of error messages, complex algorithms, and even supported web browsers, screen sizes, etc. While his software team may not need to include that much detail in their Functional Specifications (because they have experience with safety systems and they seem to rely much on non-written communication), Greg felt that the Functional Specifications of the MPC would provide a good example for his team and could serve as a general software development guide. He recognized that every organization is different and intends to review this with Vernon before meeting with his group on this. Furthermore, he needed to clarify the legal aspects of using this MPC document before doing anything with it.

It was 10:00 a.m. and Greg phoned, since there was still time to reach Ray in his office in the UK. "Villaloberg speaking," uttered Ray.

"Ray, this is Greg, sorry to bother you, but during our previous conversation I forgot to mention a possible legal issue regarding the use of the CAISTOS Multivariable Controller documentation and software."

"Legal issue?" Ray responded and carried on "hell, you and I defined the unit. Did you sign a secrecy agreement at SONARES?"

"No there is no difficulty on my part, but remember CAISTOS paid for all the documentation including the software development effort by COBOS," said Greg.

"This should not be any problem as long as you make the MPC available as a standard option on your new product offering. After all, they decided not to implement it, but once it is proven, I am certain that most of these production platform processes would want to use it. Anyway, I will contact their legal department at headquarters and let you know their response via e-mail," said Ray.

"Thanks Ray," replied Greg and they hung up.

Four days later Greg receives the e-mail response from Ray:

Hello Greg,
We have received clearance from CAISTOS legal department for your company to use the MPC software

and its documentation – as contained in CAISTOS Purchase Order XX-XXXX to SONARES.

They have reviewed our request and state: Permission to copy, modify, and distribute this software and its documentation, with or without modification, for any purpose and without fee or royalty is hereby granted, provided that the System Supplier which utilizes the software and documentation or portions thereof, for development of a product makes such product available to CAISTOS at the standard prices and purchasing conditions.

Disclaimer – This software and documentation is provided "AS IS," and CAISTOS makes no representations or warranties, express or implied.

Regards, Ray Villaloberg

Greg was going to ask Ray to forward a copy of the letter from the CAISTOS legal department and wait until he received the duplicate before talking to Vernon about developments on the AROBCO Esmix bid and the potential of AROBCO purchasing a prototype for an MPC trial installation.

And then coincidentally, just after receiving the e-mail from Ray, he heard Vernon shouting "Greg, can you please come to my office? I have Jim on the phone".

He rushed to Vernon's office. "Hello, Jim. I have Greg here now. Can you please repeat what you heard at AROBCO," said Vernon.

"Hello, Greg. I just came out of a Safety System maintenance meeting and the maintenance supervisor mentioned that his new boss, Ray Villaloberg, is talking with you about a trial installation of our new system."

"Yes, Jim, I told Vernon a few days ago that I spoke with a former colleague. He works now at AROBCO. I was going to get 'my act together', I mean check the probability of us being able to deliver, before presenting the potential opportunity to Vernon. By the way, Ray Villaloberg asked me to call him in about two weeks. Nothing has been promised to AROBCO," said Greg.

"OK guys, things seem to coming together all at once here. Greg, can you get with me in a couple of days, or whenever you have 'your act together', to talk about the AROBCO Esmix opportunities."

"Sure, no problem, and good-bye Jim," said Greg.

Vernon wished Jim success on another potential project and ended the conference call; he then turned to Greg and said, "actually let's talk about where we stand on R&D over lunch on Friday." "OK," replied Greg and walked back to his office.

Greg knows that from a perspective of development progress, his challenges are the software tasks of the Safety System enhancement projects. On the new controller, Ken is reporting his progress almost on an hourly basis. They are in constant communication about data entry details, function details, etc., and that is expected to last another four to six weeks.

On hardware, Paul is informing him almost daily on the advancements. But he does not really know where the software development tasks of Peter, Richard, Donna and Dave stand. He is not cognizant of the status of the new sequence of event recorder (SOE) development, the alarm management enhancement, the shift report generation and the new process value trending.

So far he has been focusing on the definition of the new controller and wanted to avoid confronting difficult situations with the existing Safety System developments. In the back of his mind, Greg is concerned that these projects will not communicate schedule overruns until late in the project when corrections are much more difficult and the consequences much more severe. From talking to Peter, Greg knows that the software team is often experiencing problems in meeting schedules. For them, this situation is no picnic either as Peter emphasizes. By the time they have missed repeated deadlines, their credibility is lacking and the people could get 'burned out'. Greg is worried.

Greg's experience over the past years has been that the most common motivation for implementing a measurement program is to track progress. Of course, the effective use of any progress measure requires an honest desire from each member of his software team to know the real status of their project and a willingness to take action to correct problems. Greg does not really know his people yet. But he is determined to change that.

He will talk with them more frequently and put together a table to measure the progress as a percentage of project activities that have been completed. He begins with an easy-to-implement progress measure, one requiring only planned start and end dates for each major activity, along with periodic estimates of the percent of each activity complete. Biweekly, a percent complete estimate would be provided, based on the project programmer's assessment of how much was actually accomplished up to that point. This report is used to convey progress and current status at the R&D staff meetings.

Software Progress Report

Phase	Start Date	Planned End Date	Percent Complete
Document Business Requirements	X/XX/2016	X/XX/2016	Adequate?
Document Technical Requirements	X/XX/2016	X/XX/2016	XXX
Develop Design	X/XX/2016	X/XX/2016	XX
Code and Unit Test	X/XX/2016	X/XX/2016	XX
System Test	X/XX/2016	X/XX/2016	XX
Training	X/XX/2016	X/XX/2016	XX
Data Conversion	X/XX/2016	X/XX/2016	XX
Installation	X/XX/2016	X/XX/2016	XX

Greg is convinced that these progress reports would reinforce peer review and perhaps even encourage quality improvement. But how to convey to the programmers that these reports are not intended to be a burdensome exercise in documentation, but rather intended to aid evaluation required for project scheduling? And, that the importance of progress reports is noteworthy as they serve as a means of communication for corrective activity. He feels that an effective two-way communication with each programmer, shared in an environment of trust, may give him a chance of getting the message across. He takes the report format and goes over it with each of the four programmers, excluding Dave at this time, trying to make them feel connected and important. It seemed to work as they showed understanding and promised Greg to have the reports ready before the staff meetings.

Greg is, however, aware that due to the present increased project activity, multitasking kills efficiency in his team. The more work items that are in progress at any given time, the more context switching, which hinders their path to completion. That's why a key target is to manage the amount of software work in progress. Work-in-progress supervision highlights bottlenecks in the team's progress due to lack of focus, people, or skill sets.

Greg is fully conscious of the present struggle within the team's challenge of getting things done. He feels some of the battle could be won by changing their mindset. Many people never take the time to define their top priorities. What's my most important challenge? Greg has always been convinced about efficiency through focus. He often asks himself – 'What is it that only I can do well? What are the core competencies that my company needs to focus on to be profitable and grow?' Focus has been the key to his success so far.

Prior to the luncheon meeting with Vernon, Greg made copies of the Software Progress Report, which each programmer submitted and which they discussed at the R&D staff meeting. He also made a replica of the Advanced Control Wizard (ACW) Overview. He intended to hand the information to Vernon but found him not to be in his office. He asked Tina Alexander, Vernon's secretary, where he was.

Tina replied "Greg, Vernon is in a meeting with his financial advisor. Hopefully, he makes it back before noon; I already made reservations at the Italian restaurant for you two." Greg put the copies on Vernon's desk and went back to his office.

About twenty minutes past 12:00 p.m., Vernon glanced into Greg's office and said "Sorry to be late Greg, ready for lunch?" "Yes, of course" Greg replied.

As they sat down in Vernon's car, Vernon took a deep breath and said "Wow, what a week this was and it's not over yet. Greg, I need a good meal and some virtuous news." Since it was only a few minutes to the restaurant there was no time to talk before they arrived at the door. Vernon and Greg got the usual pleasant greeting from the owner and Vernon's table was ready for them.

They were seated and in place of Vernon's usual chit-chat about friends and family at the beginning of the meal, Vernon said, "Looks like things are going well for you Greg. Paul and Peter have only positive stuff to mention about you, and by the way, I like your software progress reporting, it breaks down each task activity."

"Well, everybody is trying to do his best in spite of present overload struggle" replied Greg.

"Your encouragements seem to make a difference though" said Vernon.

They ordered the meal and Vernon continued, "I took a look at your definition of the Advanced Control Wizard."

"It's only an overview," Greg interrupted.

"It seems as if you have it all figured out" said Vernon. "I want to know everything about it."

"Yeah, it is exciting. The whole thing progressed much faster than I would have forecasted. It was mostly due to my former customer, Ray Villaloberg, and his recent joining of AROBCO. We worked together at the CAISTOS production platform project and defined a Multivariable Process Controller for the compression and separation process. Well, I did most of the specification work but I must say that Ray had many inputs on some particulars of the process. He is a very detail-oriented engineer."

"Jim told me, though, that he is in a key management position at AROBCO," Vernon interrupted.

"He just joined them, I am not sure which position he occupies there," said Greg and continued, "anyway, Ray is still interested in pursuing the Multivariable Process Controller application, and there is an opportunity for us to get all the documentation and software listing of the development effort that was expanded at CAISTOS. The software listing for the MPC is more than three times the size of our whole Safety

System software package, no exaggeration! Here is the e-mail I received from Ray." Greg pulled out a copy from his jacket and showed it to Vernon.

Vernon carefully read the e-mail about the clearance from CAISTOS legal department for the use of the MPC software and its documentation and asked, "does this mean that they would release the software free of charge?" "Yes, CAISTOS paid for all the development. For us it would involve porting the software to our platform. It is now on UNIX, and it would be a significant verification and testing effort - several months of work. However, I am reasonably confident that we would get reimbursed for it," said Greg and carried on, "I will get an estimate from Ken on the software portion next week, if you agree on pursuing that opportunity."

"This is almost too good to be true. Of course, I approve to chase this opening for our new product. Do I need to do anything at this time?" asked Vernon.

"No, I will call Ray Villaloberg in a couple of weeks and tell him that we are interested. Hopefully he has the necessary influence there at AROBCO to persuade them to issue a PO to us."

Vernon did not order any dessert, which is unusual for him. As they left, Greg sensed some uneasiness and asked him, "Is the big Safety System job coming along alright?"

"Yes, we are making substantial progress. Unfortunately, my financial advisor and main investor think such a contract award means immediate dollars into our bank account. He is questioning the terms and conditions of the purchase order.

He wants me to re-negotiate the order with AROBCO. Don't worry Greg, it will be OK," said Vernon and they drove back to the office.

Back at the office Vernon asked to see the CAISTOS documentation package. When Greg put the three inch folder on his desk and opened it to show him the software listing, he said, "wow! - You were really not exaggerating. You should have Tina make a copy, not for me, but in case you give the documents and listing to Ken, you may want to keep a duplicate."

"Yes, I was going to do that, but if Tina has time that helps," said Greg.

"Sure she can do that. Have a good weekend, Greg. I need to go."

"You too Vernon, and thanks for the lunch," replied Greg.

As usual, during the weekend Ken called to go over some function details. Although the call often lasted several hours, it was rewarding for Greg because he could see how much progress Ken was making and felt a real sense of accomplishment in being able to help him do so. At the end of this Saturdays' conversation, Ken asked Greg if, for the upcoming week, he could schedule a day for a software walkthrough since he is finished with the configuration structure and the basic functions.

There have been plenty of ups and downs during Greg's first weeks on the job, but being able to tell Vernon that Ken has completed the basic coding for the new controller is something Greg is really looking forward to. Greg felt really great. He called his girlfriend and asked her if she wanted to go to the beach and that he would reserve a hotel for tonight. They had spent little time together during the past few months, and Greg was so glad that she had understanding for his long hours at the office and for his continuing to work at home.

Monday afternoon, as Greg was working away on his Application Guide, John Kramo, the Project Manager, came busting into his office and said, "Greg, your team is holding up the CAISTOS project. What do you plan to do about these software schedule delays?"

Greg, with a baffled look on his face responded "What happened? Peter, Richard and Dave have been working on the tasks of the Safety System project only. Nothing has changed. There is no other assignment for them. So what are you telling me? What occurred all of a sudden?"

"That is what I was asking them in this morning's project meeting when they came forward with different completion dates" said John and continued, "I am not going to accept this. Unfortunately, Vernon is not here, otherwise the three of us would have had a serious talk."

"OK John, there must be a misunderstanding. I will get together with my team tomorrow morning and will get back with you afterwards. Is this all right with you?" replied Greg.

"OK, tomorrow then," said John and he left Greg's office.

Greg leaned back on his chair, took a deep breath and whispered to himself 'here we go with that 90 percent done syndrome in software status reporting'. He suspects that, whether through wishful thinking or general optimism, his people may not have given realistic dates to John. Or perhaps they have not taken software test time into consideration, and he doubts that John has asked the right questions. He was certainly surprised by John's accusatory tone and his threat to immediately involve Vernon. 'Considering that John was the overly friendly one during his first days in the office, he certainly had a short honeymoon period with him' thought Greg. These are the typical moves of a schemer. 'The sooner I resolve this, the better off I will be perceived in the work environment with John and his project team.

Greg could not get John's angry behavior out of his mind when he got home and found it difficult to relax. He knows from years of experience that offices are competitive places and not all of the co-workers will have your best interests at heart. The reality is that there is likely to be someone who will try to build himself up at your expense and will see a perceived schedule delay as an opportunity. This can come in the form of open or behind-the-back comments. In other cases it can be in the form of someone pretending to be your friend while sabotaging you on the job and trying to grind down your confidence. He says to himself, 'the trick here is to not give the schemer any traction'. Do not come into the office tomorrow looking weak and like a victim.

Have the attitude, 'It happened and I'm going to figure out how to cope with it. Fix the problem!' The vast majority of people want you to succeed, and so far it has been a great experience at MICGEN. 'Focus on those positive interactions to build your confidence and avoid the office schemers by not acting like a victim' reemphasized Greg to himself and he finally calmed down.

Next morning, as it frequently was the case, he ran into Peter at the coffee corner. He asked him "Anything special happened yesterday at the project status meeting?"

"Yes, John was all upset. After Paul told him that he did not approve of some of the alternate component selection for delivery time enhancements; he then asked me if the software was going to be completed on time. I responded that we have to have hardware for the final tests. He blew his stack and yelled at me declaring 'you gave me software completion dates - are they all of a sudden no good?' Peter said.

"Well, it was probably a misunderstanding," Greg said. "No, I don't think so since I told John several weeks ago that the software dates did not include the test time. John has a tendency for overoptimistic estimates and then blames others when delays occur. Right now, he is all bent out of shape. Components, cabinets, everything seems to experience delays. Hope he doesn't end up back in the hospital." responded Peter.

"Well, have a good day, Peter" said Greg and he went straight to John's office. John was just marking up things on his blackboard as Greg stopped at his door. "Hello John, it appears that there was a misunderstanding yesterday. I just spoke with Peter," said Greg. "Yes, you may have a point. I thought about it yesterday evening. Anyway, my schedule is a mess," replied John and turned back to his Blackboard. Greg thought it is better not to make any other comment to somebody in such a bad mood and went back to his office.

Later on in the morning Ken left a message on Greg's phone, asking him if he would be available in the next few days for a software walkthrough. Greg could see from the callback number that Ken phoned from his apartment and called him back. "Hi Ken, when to you want to do the review?"

"Would Thursday morning be OK with you?" Ken replied. "Sure," said Greg, "until Thursday then." "It will most likely take all day," said Ken. "Yeah, hopefully it does and we can not only verify the data entry but also check out some of the functions," said Greg. "Oh yes, I am sure we will be able to do so. I have already tested some of the basic functions," said Ken.

"OK, then let's go after it on Thursday. You have a good day," responded Greg.

Ken was ready to show Greg that his software is working well. It was much more than the structured walkthrough Greg expected. When they met Thursday morning, Ken had his software downloaded to the emulator board and was changing parameters on the Constraint function.

"Hello, Ken, it appears everything is at present running on the simulator?" said Greg.

135

"Yes, it is," said Ken with a proud smile on his face." And so far, no errors. I have been here for over an hour, where have you been? It is already 7:00 a.m. Are you ready to start?" Ken joked.

They tested the PID and twenty-three other functions, including some advanced functions such as Conditional Fallback and Constraint Limit. After several hours of tests they found only a few minor errors.

"Ken, this is great," said Greg. "I realize that you can't test the field I/O until the prototype hardware is available, but what you have been able to accomplish in that short timeframe is outright fantastic."

"I cheated a bit. I was already working on some of it before I came on board. Remember the phone calls?" said Ken.

"We have to show this to Vernon next week" said Greg.

"So what is our next priority?" Ken asked.

"Well, we were going to continue implementing our control functions, and as you know, I have been working on the Application Guide," Greg said. "But something unexpected came up last week. I have an enormous challenge for you Ken. Remember, we were talking about including advanced process control functions in the future. Well, let's go to my office to discuss this," and both got up and went to Greg's office.

"Here is an overview of what I call the Advanced Control Wizard module." Greg handed Ken the definition he had written a couple of days ago. "It describes a software package that includes the Multivariable Process Control, the Constraint

Limit Control and a Process Configuration Module. It would, of course, also need a set of desktop tools for control analysis and design. While Multivariable Process Control, the heart of the Wizard package, is now already installed in many applications, only the complete Advanced Control Wizard Platform will deliver significant process improvements," Greg said and continued, "Ken, this would truly be a breakthrough in advanced process control."

"I understand that assessing the quality of software - especially someone else's - is a tricky balance between objectivity and the very subjective, but in your case very valid, individual user experience," said Greg, and he handed Ken the Multivariable Process Control documentation - a three inch thick folder. Ken had a look of surprise, as if he'd just swallowed an ice cube. Greg continued, "Ken, I would like you to make a quantitative assessment of this software package in terms of sustainability, maintainability, and usability and give me your estimate as to how long it would take you, in man-months, to implement it on our platform, if you determine that this package should be and can be used. This software package relates to Multivariable Process Control only. It has been released as source to a client. We probably can obtain it at no cost. The question is 'should we?' Greg paused before continuing, "how long would it take you to make a preliminary assessment?"

"Wow!" said Ken, as he flipped through the software listing contained in the folder. "That is a big program, but I should be able to get back to you by tomorrow evening." "No, I don't want that, Ken. Don't put too much pressure on yourself. Let me know sometime next week," said Greg.

Tuesday evening Greg receives a phone call from Ken. "Hello, Greg, I have summarized my valuation of the software. I will give it to you tomorrow morning. It is several pages. Since there are more than a dozen issues, I have a recommendation at the outset listing of what I believe to be the most pressing issues encountered, along with a rough estimate as to the impact of addressing these, e.g. time to do, impact upon architecture, etc. Overall, the software is well-structured and of good quality. We certainly could use most of it. My estimate is that it would take about one month of porting it to our OS platform and then about another two to three months of integration and test. It is very well documented, thus the manual generation would be a minor effort."

"That sounds really good Ken, thanks," said Greg.

On Wednesday afternoon, Greg went with bid copies in hand to Vernon's office. "Hi, Vernon. Could you spare a moment to review the proposal for the AROBCO Advanced Control Wizard?"

Greg went through the budgetary pricing, the cover letter, and Ken's report on the software assessments, etc. They spent about an hour discussing various parts. Then Vernon remarked, "Good; this is very comprehensive. Let's go for it." "OK, I will call Ray Villaloberg tomorrow morning," Greg replied.

MICWIZ System with Advanced Control Wizard (ACW) – Budgetary Pricing

Item	Description	Qty.
1	**Advanced Control Wizard (ACW) Module** Model: ACW: A2-B1-C1-D1-E2-FX – TMR redundant 20 AI, 8 AO, 24 DI, 12 DO - 64 I/O points on I-Safe Intelligent Termination Panel Standard Functions plus ACW Firmware: - MPC - Multivariable Process Control - CLC - Constraint Limit Control - PCG - Process Configuration Genius	1
2	**Workbench Wizard Workstation** Model: WWW: A1-B1-C1-D1-EX Type: Desktop High-speed IBM compatible PC: - w/ 17" LCD monitor/keyboard/mouse - WWW System and Application Software Module - Historical Storage and SOE - OPC Server - ACW Desktop Software Tool set	1 lot
3	**Review of Existing Compression & Separation System** Estimated Time: 5 days field "walk-down", 5 days analysis	1 lot
4	**System Design and Configuration** (excluding cost of field trips, if required)	1 lot
5	**Dynamic Process Simulation** (customer to provide field data within 2 wks ARO)	1 lot
6	**Factory Acceptance Test (FAT)** – 5 days (at MICGEN)	1 lot
7	**Site Acceptance Test (SAT)** – 5 days (excluding travel and living expenses)	1 lot
8	**Commissioning & Startup Assistance –** 2 Eng.10 days (excluding travel and living expenses)	1 lot
9	**Site Training** – 10 days (excluding travel and living expenses) **Total System Net Price**	1 lot

Attachments:
- Quotation Cover Letter
- Legal clearance requirements of MPC software and documentation from CAISTOS
- MICWIZ Advanced Control Wizard (ACW) Overview
- MICWIZ Process Application Guide
- MICGEN Standard Terms and Conditions
- MICGEN Standard Site Service Support Rates

Delivery: Nine (9) Months ARO

Payment Terms: 30% upon receipt of order; 40% upon shipment from factory; 30% completion of deliverables.

Next morning Greg phones Ray in his office in the UK and Ray picked up after the first ring with, "Yes, this is Ray Villaloberg speaking."

"Ray, this is Greg, how are you? You asked me to call you back on the Multivariable Process Controller. We have analyzed the application, and I want to express that we are interested in pursuing this," said Greg.

"Great, can you please send me a proposal by e-mail?" replied Ray.

"Yes, of course, I have already prepared a budgetary bid and can forward it to you right now, is that OK?" responded Greg.

"Yes, please do so. I hope you made it not too budgetary. I need sufficient details to be able to present it to my boss," said Ray.

"I believe that you will find it quite comprehensive. Please call me on any comments you may have," replied Greg.

"Will do. Are you going to be in the office for the next couple of hours?" asked Ray.

"Yes, I will be in all day. Should I not be at my desk, please have me paged."

"OK, you will hear from me soon; till later Greg," said Ray and he hung up.

As Greg scans in the proposal document and e-mail's them to Ray, Paul Bingham knocks on his office door. He normally leaves it open, but during client phone calls, Greg usually closes the door.

"Hi Greg, would you have time to go over the latest controller board layout with me? We have been able to accommodate the increased memory capacity and have also added a new communication chip," said Paul.

"That is super," said Greg. "Can we go over the layout here in my office? I am expecting an overseas phone call from a customer."

"Yes, let me get the drawings," said Paul, and he returned with a stack of documents. He could see by Greg's facial expression that he does not expect to review the whole pile of papers and said, "There are only two drawings I would like to review with you. The others are for my reference."

"OK, let's do it," said Greg.

Paul explained the main circuits and components to Greg and the noise immunity provisions of the electronic design, especially in the process I/O circuitry. He then covered specifics on the processor, the communications and the memory management. And as he went over the specific aspects, he repeatedly emphasized the advantage of the single board architecture and of the TMR reliability.

"I am impressed" said Greg. "And how long do you think it will take to get a couple of prototypes?"

"I have checked component delivery times and have a listing of alternate components for most of the chips. I would estimate about four months. Although the board manufacturer we normally use quoted only three months, there are usually some delays with the military spec I/O components" responded Paul.

"OK, I appreciate you keeping me up to date on all this. Thanks Paul," said Greg.

It was not five minutes after Paul left Greg's office that Ray called back. Greg saw Ray's number on the phone and answered, "hi Ray, I hope that I included all the information you expected in the bid."

"Yes, it looks very good. I do think that your estimates on commissioning and startup assistances are low. But, since you included the site service support rates, I can make the adjustments there. Ah, and I also will check on the training. We may have to double this for the reason that we would have two separate groups. Again, I can take care of this.

Oh yes, I almost forgot, your proposal describes Constraint Limit Control and a Process Configuration Genius. Do these functions exist?" questioned Ray. "Also regarding delivery, your quote states nine months. This is getting into our new fiscal year. Is there any way you can do better?"

"Let me address the Constraint Limit Control and the Process Configuration first. These functions already exist in a simplified form in our new system. Concerning delivery, there is a high probability that we can deliver in seven months, but as you know Ray, there are often unforeseen delays. I would feel better if we leave the delivery at nine months," responded Greg.

"OK, Greg, I hope that I can get this approved. Well, maybe I can absorb it in my maintenance budget. Will think about it. I should be able to let you know sometime next week how the situation here develops," said Ray.

"I look forward to hearing from you next week then," said Greg and they ended the conversation.

With the high potential for the Advanced Control Wizard, Greg needed to focus on the Application Guide for typical production platform compression and separation processes. It was imperative that he provide Ken with the information for function testing purposes. Thus, after the telephone conversation with Ray, he continued his work with the office door closed. The concentration on the application task did not last very long as he received a call from Tina Alexander. "Greg, did you forget about the health insurance meeting?" she said.

"Oh, I am sorry. I will be right there," responded Greg and he hurried to the conference room. There was standing room only; everybody, but Vernon, seemed to be there. Mike Jacksens, MICGEN's CFO opened the meeting with an overview of the difference in insurance coverage. The new insurance provider then gave his PowerPoint presentation. The whole arrangement, including questions and answers, lasted over an hour and Greg kept looking at his watch.

As everyone left the conference room, Peter moved over to Greg and remarked, "we are re-negotiating constantly here, what a waste of time. Could you spare a moment Greg?"

"Yes, of course. Let's go to my office," replied Greg. And there Peter said, "I heard rumors that we are already going ahead with the component purchase for the new controller and would like to bring to your attention that there is a flurry of renegotiations going on with vendors to get the reduced prices. They brought in all the vendors that offered estimates for the Safety System project and the new controller, and told them we needed to go back and revisit pricing. I understand that they came in a few percent less than they were originally, but it has caused havoc with alternate component specs and is now resulting in delivery problems."

"Oh, is that one of the reasons for John Kramo's project schedule issues?" asked Greg.

"Yes, perhaps the main cause. But the real concern should be quality. With all the substitute parts use, we could end up with inferior products," replied Peter. "Al Murrell, our purchasing guy, sometimes articulates that most contracts are no longer set in stone and is renegotiating with vendors, landlords, insurance providers, and the like. While on the surface this may save a few dollars, in reality once delivery and quality issues are considered, this ongoing negotiation trend can put us, in my opinion, in a grave situation; especially now with the AROBCO project."

"Who is behind this? Is it just Al Murrell trying to save some money?" asked Greg.

"No, actually Al often makes remarks that this renegotiating work is causing him much additional effort. It is Mike Jacksens, our CFO, and David Freetman that are behind all these constant negotiations" said Peter.

"David Freetman? Who is he?" asked Greg.

"You don't know Freetman; he is our principal investor" said Peter. "He seems to be involved in everything, even office supplies."

"Thanks for telling me all of that. I will keep it confidential," said Greg.

"You are welcome," replied Peter and he walked back to his cubicle.

Greg leaned back in his chair, took a deep breath and said to himself, 'I am so focused on that Application Guide and the new controller that I am not aware of major internal company problems'. He thinks about Vernon's comment, after Friday's lunch, on the terms and conditions renegotiations with AROBCO. He knows that re-negotiating fixed contracts is never easy and can potentially sour business relationships, since there are few deals that are mutually beneficial. Before being able to genuinely renegotiate terms one must assess the effect that such an action may have on future orders from the client. In AROBCO's case this could threaten the survival of MICGEN. 'Has David Freetman considered this side of the equation?' Greg asked himself. He wants to talk with Vernon and goes to his office only to find out from Tina that Vernon is downtown in a conference with David Freetman. He asked her to tell Vernon that he would like to talk with him when he returns from the downtown meeting.

Late in the afternoon, Vernon glimpsed into Greg's office and said, "Greg, I understand that you wanted to see me."

"Yes, Vernon, can you spare a minute?" asked Greg.

"Sure, anytime," replied Vernon.

"Let me get directly to the point," said Greg and continued, "I am concerned about our situation with the CAISTOS Onshore Safety System contract and it does not relate to tasks of my team. So you may say or think that 'this is none of your business Greg', but it is in reference to potential delays due to renegotiations. Pricing debates about component alternatives probably will affect delivery and may also influence quality. And, last Friday you mentioned to me something about renegotiating the terms and conditions with AROBCO. All this has me worried because the bottom line is while renegotiation can be a strategy to cut component costs and change payment terms, in the case of the AROBCO, it has the potential of affecting the offshore production platform new controller trial installation and also the new control system bid. Hopefully everyone understands that we are in trouble if we upset the business relationship with our key customer, and hopefully Ray Villaloberg does not get wind of this."

"I fully agree with you. We are taking high risks and I am fighting David Freetman on this. Believe me, I am aware that our relationship with AROBCO is worth more than the risks we take with these renegotiations, but I have a difficult time conveying this to David," said Vernon and continued, "when it comes to money, David Freetman is a tough guy to deal with. However, so far he has always come around and I believe that we can sort it out this time as well."

"Sorry to have bothered you with this, Vernon," said Greg.

"That's OK, Greg. Believe me, I appreciate that you express your concerns to me directly. Don't hesitate to call on me," said Vernon and he left Greg's office.

Greg was ready to leave the office when Monica came in with a handful of papers that looked like brochures, saying, "hi Greg. I know it's late but could you spare a few minutes?"

"Sure Monica, what have you got there?" said Greg pointing at the papers in her hand. "Well, this is your brochure," she said and put the top leaflet on Greg's desk.

"Wow, where did you get that picture of our new controller from? That looks great," said Greg.

"We created it from Paul's manufacturing drawings. Doesn't it look real?" asked Monica.

"It's hard to believe what can be done with today's graphic design tools," commented Greg and stared open-mouthed at the cover, like a fish out of water.

"We did not change the content of the brochure, just the photos," said Monica. "Here is the updated company brochure. I increased it to eight pages. I believe that the information now flows in the right order. I analyzed the competition's literature and also feel that from a reader's point of view we have a good format now."

"Let me take about ten minutes to look that over," said Greg. "Sure, I am going to get myself a cup of coffee, can I get you something too?" asked Monica.

"No, thanks," replied Greg, as he stares at the front page and then scrutinizes the other pages.

Monica kept the company message warm and emotional, and she focused on providing a solution for the reader. She attractively listed the benefits of the solution. The brochure answered the questions in a logical sequence following the typical safety and control system business reader's train of thought. The front cover appearance was also very professional – photos and all, with thought-provoking statements that motivated a reader to pick up the brochure and open it. Greg was impressed, and as Monica returns, he said, "my compliments, this is a first-class piece of literature."

"Thanks, Greg, I have spent many hours and several weekends trying to understand the nature of our new business, the questions readers will have and what they will want answered before they will consider looking seriously at our company. I have applied that to our company profile leaflet, to the company PowerPoint presentation and to our website," said Monica.

"I am impressed," said Greg and continued, "now let's convince Jim to invest in professional printing. Yes, office printers do a great job. But they aren't as good as real printing, and a reader can tell the difference. We need to select a heavyweight paper that feels substantial in the reader's hands."

"I agree Greg. Anyway, printing costs are lower than they have ever been," said Monica.

"OK, this literature is superb. You have a good evening," said Greg.

"Thanks," said Monica and she left.

Greg did not hear from Ray Villaloberg and it was already Friday morning. He was getting concerned. As he was about to leave for lunch, an e-mail from Ray came in advising that the order will be placed by purchasing early next week and asking Greg to give him a phone call around 7:00 p.m. Ray's time. Greg took a quick lunch and returned to the office to call Ray at home. Nobody picked up, so it switched to the answering machine. Greg left a message that he will call back in ten minutes.

When he tried the second time, Ray picked up immediately, answering, "hello, Greg. Sorry I missed your call, but I just stepped outside five minutes to talk to the neighbor next door. Anyway, I need to talk with you about the order."

"Please, go ahead," replied Greg. "First, I would prefer that they forward the order directly to you instead to your purchasing department, so if any questions arise on the PO, you are the straight-line contact; is that OK?" asked Ray.

"Sure, no problem" answered Greg.

"Second, purchasing changed the T & C's from yours to theirs. They advised me that your company is familiar with our terms. Ah, and we left the delivery and the payment terms as per your bid," said Ray.

"Yes, considering the number of Safety System projects we delivered to you, we should be familiar with your terms and condition requirements. I will check and call you back in case of any concerns," responded Greg.

Ray continued, "Thirdly, the total price reflects the increase in field service and training hours; I have changed the Site Acceptance Test to 10 days, the field assistance to 30 days and the training to 20 days. I realize that the MPC is not fully proven, but this should give us plenty of time. I don't want any change orders messing up my maintenance budget, do you hear?"

"Yes, I understand Ray," replied Greg.

"OK, you should receive the order with the drawing package, the same you most likely already have, and the disk with the source code next week. Our purchasing department will send you an e-mail advising the Fed-Ex number either Monday or Tuesday," said Ray.

"Superb! Thank you very much, Ray," said Greg. "I am counting on you and look forward to working with you again," said Ray.

"My pleasure, and thanks again," replied Greg and they hung up.

Greg went right away to Vernon's office to give him the good news. "Vernon, I just got off the phone with Ray Villaloberg. We will receive the PO for the new system next week," announced Greg.

"This is great!" replied Vernon with enthusiasm in his voice. "I did not expect to receive the order without them being convinced that the controller exists and had been at least factory tested. This is still hard to believe even though you indicated a few days ago that they will place a PO. Your buddy Ray must have a lot of stroke."

"There are a few things I need to verify with you. They changed the T's & C's from ours to theirs" said Greg. "OK, I expected that. How about the payment terms?" asked Vernon.

"They left the payment terms and the delivery as we quoted," replied Greg. "Ray also added 35 days in field services - this should improve our profit margin to over 50%, and when you take into account that we can absorb all of Ken's development time and also Paul's and Ken's test time of the new product, I hope that you will permit me to make Paul's efforts in producing the hardware a top priority," Greg stressed.

"Yes, you are his boss. But I know what you mean, I will make John Kramo aware of it," replied Vernon.

"Oh and I almost forgot, Ray made me promise that there will not be any change orders on this job."

"Well, it is your project," replied Vernon.

"Sorry, Vernon, I said that in consideration of what we recently talked about, I mean renegotiations," said Greg.

"OK, you got a point. I will make sure that Al Murrell gets his purchasing instructions for the parts of the new controller directly from you or Paul," responded Vernon.

"That will decrease delivery surprises. Thanks Vernon," said Greg.

"Well Greg, if you don't have any plans for tonight, I suggest that we have a drink or two to celebrate this; actually, let's go now. It is almost 5:00 p.m." said Vernon.

They went to the Italian place again, Vernon's favorite hangout. They talked about their past and Vernon revealed his challenges with David Freetman, asserting, "I have been fighting this battle for a year-and-a-half now. Your project should help, but I don't know if this guy will ever be satisfied."

On Monday morning Greg received the e-mail from AROBCO purchasing, informing him that the purchase order is expected to be delivered by FedEx on Tuesday or Wednesday and giving him the PO number and FedEx tracking number. The attachment included a PDF copy of the cover letter and the budgetary pricing sheet with the changed site man-days and the revised dollar total. 'That is the way to start a day' utters Greg to himself. It is one of those days where everything seems to be falling into place. Paul showed him the quotation update from the board-stuffing company stating that they could fit five prototypes into their schedule almost immediately. Ken came into his office telling him that he tested the Constraint Limit function, and even John Kramo came by his office to tell him that the Safety System job schedule is in reasonable shape again. Greg was not quite sure why John informed him, but maybe this was just one of those unusual Mondays. And when he walked into Vernon's office, here was Vernon with a mankind-loving look, asking, "are you bringing me good news too, Greg?" "Yes, I am. Here is the e-mail from AROBCO with the PO number and the revised total of $379,200," said Greg.

"Well, we better get that show on the road then. Tell Paul to order the prototypes and tell him that if Al or Mike give him a hard time on pricing to see me," responded Vernon.

"Thanks Vernon," Greg said and returned to his office thinking that considering the start of this day, maybe some magician finished his Application Guide. Of course, when he opened the document, it was just as he left it last Thursday. Anyway, he needed to change his priorities to the tasks of the new system PO, so Greg spent the rest of his day re-familiarizing himself with the MPC documents.

On Tuesday around 10:00 AM, the FedEx package with all of the PO documents and the source program disk was on Greg's desk. He compared the documents in the package to the one already in his possession and found that some of them where of an older revision. His concerns that AROBCO sent an outdated documentation package were, however, minimized when he found that the software listing revisions matched. He remembered well that he was one of the last engineering contractors leaving the jobsite at CAISTOS, and most likely all the updates had not been communicated to the client. Therefore, he decided not to contact Ray for clarification. Everything appeared complete. So, he acknowledged receipt of the package and acceptance of the purchase order to AROBCO's purchasing department.

He took the package, except for the source disk, to Al Murrell and asked him to enter the purchase order, but with a six months delivery date instead of the nine months shown on the order. "What do I need to do with it except for the log-in of the PO?" asked Al.

"Paul will get with you on the equipment tomorrow," replied Greg. "Regarding component selection, we go with Paul's specifications. No alternates, OK."

He then went to Ken's cubicle and told him "full speed ahead on AROBCO. We got the order, here is the source code" and handed Ken the disk. "I am already working on it," replied Ken. "Would Friday be a good time for us to sit down and go over the function details?" asked Greg.

"Yeah, just to make sure that we are on the same wavelength," said Ken.

"Let's say around 9:00 a.m." said Greg.

His final stop on this purchase order entry tour was Paul. "Paul, you presented the quotation update from the board-stuffing company to me. Well, we just received the official PO for the new controller from AROBCO. When would you be able to release the order for the prototypes to them?"

"I was waiting until Al gets back to me on the alternate component pricing. He thinks that we can save 6-8%, but I am concerned about specification tolerances, delivery, supplier quality, etc.," said Paul.

"I just told Al to go by your specs, no alternates."

"Super! That will save a lot of hassle. Thank you Greg."

"If Al or Mike gives you a hard time, Vernon asks to see him," said Greg.

Greg ended these two days with a special sense of accomplishment. When he returned to the office after talking with Paul, he relaxed. He felt that his skills and his ability to work well with his group contributed to the company significantly. His good customer relations are paying off. And his long working hours do result in productivity. He does not believe the surveys that indicate work above 40 hours a week makes one less productive.

Greg naturally wakes up early, between 4:30 and 5:00 most mornings. He uses the time to read an overview of the news on his PC, clear out his inbox, update his calendar, and do any quick work tasks that he wants to get off his plate before a busy day begins. It is normally the most productive hour of his day. By 6:45 he is usually on his way to the office; this way he avoids the rush hour traffic.

The award of this AROBCO Esmix compression control project has made Greg the center of attention at MICGEN. Perhaps some of the people have not been convinced of the Vernon's strategy to expand from a Safety System products firm to a Control Solutions company. But now they see that the new control product actually sells. He made it happen. He created the product. In practice the hardware was already mostly designed. Only the software is new.

Shortly after 9:00 a.m., Mike Jacksens, the CFO asked Greg to meet with him. Mike is rumored to be a close friend of David Freetman, the majority owner of MICGEN. He wanted to know how Greg computed the profit margin on this job. Greg told Mike that he went over that with Vernon but that he would be glad to repeat the explanation of his calculations. And they spent almost one hour going over the gross profit analysis details due to the many questions of Mike. The queries indicated to Greg that Vernon had already forwarded his profit breakdown, even though Mike told him that he had no info on it. Mike seemed to be satisfied with Greg's clarification of the gross profit.

It was not only that meeting, but Mike's comments on all the other get-togethers, that convinced Greg that Mike was not really a CFO but the stereotypical image of an accountant, someone who is cautious, risk-adverse, detailed, conservative, even humorless and boring. To Greg, a CFO is a businessman first, and accountant second - somebody that focuses on strategic business issues rather than just getting the books to balance. And at MICGEN this is really only Vernon – he is the CEO and the CFO. As far as Greg is concerned, Mike is the chief accountant and the link to David Freetman.

And just before noon, David Freetman stepped into Greg's office and said, "hi Greg, it's great to finally meet you. I understand that we received a nice little project from AROBCO. Congratulations."

"Thanks," replied Greg. "There's been a few hours expended, not just by myself, but also others to make this happen."

"From what I heard, it appears to be a project that could be the foundation for expanding the company," said Freetman.

"Yes, it will provide us with the opportunity to expand into the control system business faster, but a lot of that has been started before the award of that job," responded Greg.

"Unfortunately, I have to run, but would you have time to join me for lunch, say Friday?" asked Freetman in a casual way.

"Of course, where would you like to meet?" said Greg. "How about downtown at Brennan's? It's close by my office. Let's say 1:30 p.m." said Freetman and he left Greg's office.

Greg was going to see Vernon to tell him about Freetman's invitation, but was informed by Tina that Vernon will be out until Thursday. When Greg enquired with Tina where Vernon is, she told him that he is in Boston. 'Something is not right' said Greg to himself, Vernon normally makes a point of apprising him of his trips.

Thursday morning as Greg walked by Vernon's office, Vernon called him, "Greg can you step in for a moment, and please close the door behind you." Vernon had a lonely face and said "I have recognized the need to move on with my career and have found a good next step." He continued, "exiting our working relationship is an emotional and sensitive experience. It occurred over a couple of weeks, not hours. There was a realization that things are not going to change with David Freetman and that I needed to make the change."

Greg looked puzzled and grieved, as if he can't believe what he is hearing, and said, "Vernon, for me that is a major hurdle, I joined this company mostly because of you. Where are you going?"

"I will tell you this afternoon, but for your information I am not joining a competitor. Sorry that I have to surprise you with all this. See you this afternoon, and please keep this confidential. Only you and Tina know about it," said Vernon.

Greg went to his office, set down, exhaled and breathed deeply. Too often he didn't pay attention to the signs of an impending problem. He will just have to wait until this afternoon to find out more details. Now he understands the luncheon invitation from David Freetman.

At 4:00 p.m. Vernon called and asked Greg to come to his office. He said, "I prepared a list of what is pending and what needs to be done soon. Greg, you have my home phone number. I have forwarded the calls to my cell phone. I will be available to answer questions. I have considered my feelings for the people here and came to the conclusion it is best for us to tell them that I am taking a leave of absence for personal reasons and that you are in charge until I return. I have cleared that with David this morning. I have already advised Tina of it. She is good in handling personal situations. Tina also has my new office contact information."

Greg did not expect that kind of exit from Vernon and interrupted, "Vernon, you are not saying that you are leaving right now?" "Yes, I am" Vernon said. "But let me give you some advice about David so you know what to expect." He paused and continued.

"First, David's personal considerations: Let me try to put that in perspective for you Greg; your previous boss, Sam Watering at SONARES, was a hands-off type. Well, David is the opposite, a micromanager in financial matters who insists everything must be done his way; that gets a little more difficult. Micromanaging is a serious issue, resulting from either a lack of trust or a need for control. I have been working with him for many years now and I am still not sure which one it is.

Second, David's plans for MICGEN: You need to know that he has been talking to our prime competitor about the value of this company. He knows that they are still wheeling from the loss of the CAISTOS Onshore Safety System job to us and now he is using your project to emphasize our potential in the control arena. But don't worry Greg that company is not in a financial position to make an acquisition. However, David is also talking with a large overseas company. I am not sure what the probabilities are with that firm. Again, I trust that you will keep all that confidential."

"Now let me tell you about Tina Alexander because she may be the key to your success here" continued Vernon. "I would define her this way. As an assistant to any manager, she possesses a mastery of office skills and the ability to assume responsibility without direct supervision. She displays initiative, exercises judgment, and makes decisions within the scope of authority. Also, considering that she is normally the first one to learn about many confidential developments involving the office staff and company policies through meetings, letters, in her filing, and dictation given by me, I must tell you her discretion is special. Moreover, she understood my weaknesses, and she will in no time recognize yours, and does not reveal them to others. She relieved me of office details, such as coordination of future activities, and the follow through of various projects. She is a good public relation person. I am sure that you will appreciate Tina's talents."

Then Vernon closed by saying, "let's keep in touch with one other in the future, and Greg you can be certain that you are going to remain in my network. Now I want to say good-bye." They shook hands and Vernon left. While Greg was sure that he will hear from Vernon soon and see him again, he almost felt like crying. He did not expect Vernon's rapid departure and the termination of this close professional relationship. He went to his office, took his attaché case and drove home so people could not see his sad facial expression.

Greg could not relax at home. Many things run through his head. 'Taking over a job that has been filled by a dynamic performer and good manager, such as Vernon, would certainly be challenging. The employees may have trouble accepting the change or shifting their loyalties. They may doubt his ability, or even try to sabotage his efforts'. One thing Greg was determined to do was not change his leadership style of work; he is resolute to being his own person. He certainly does not want to be seen as a cheap imitator of Vernon, but he does not want to come across overconfident either. He knows himself and is confident that he can communicate in ways that are authentic to his personality. He will look for his own strengths, his own aptitudes. And one of his strength in terms of connecting with others in difficult situations is to tell them 'Look, I am not a genius, but somehow I know that we can solve this problem'. Greg knows that people had a lot of confidence in Vernon. So when he takes over, they may become anxious because they are not sure what will happen. They will probably start worrying that, if in this whole company management change things go wrong, they might suffer.

Next day, on Friday, he was at the office at 6:00 a.m. When he sat down, he glanced at a proverb he posted on the wall that says 'If things are getting easier, perhaps you are headed downhill'. That is a sentiment Greg hangs onto and he put it there for others to see it. Then he said to himself 'It was Vernon's idea. He had a lot of confidence in me. He chose me to take over for him instead of someone else here, even though I have been here only for a short time'.

So Greg pulls himself together and thinks 'while it will not be easy to fill Vernon's shoes, I know that I can take steps to gain the trust and approval of the employees. It has worked with my R&D team'. He will work to mobilize the teams around the new mission of providing a total system solution. He knows this business well, and he believes that this will be a good basis to work on together. He certainly does not want people to look in the rear-view mirror; he wants them looking forward. Greg also knows that the main challenge facing him is fulfilling the expectations of their customers. Vernon was very well liked. Winning customer loyalty is more difficult than winning over a team, because he won't have the ongoing daily contact to make inroads. The key is to really understand what they loved about Vernon, and for that he thinks Tina can guide him. He will arrange a meeting between them and have her talk about what made Vernon a star in her eyes.

As Greg looks out the window, he sees Tina's car pulling into the parking lot. He waited ten minutes before he walked into her office and said, "Good morning, Tina." She was red eyed and saw that Greg looked into her eyes.

"Good morning, Greg, I could not sleep last night," she said.

"I did not sleep well either" Greg said and paused before he continued. "Vernon spoke highly of you, so I am confident that we can make it through that and come out of in a positive way."

"Thanks for the confidence," Tina replied and looked at the list Vernon left.

"I have a copy of the same list. Let's go through it this afternoon. Regrettably, on Wednesday I promised Ken that I would go over a number of things with him this morning. I did not expect this to happen," said Greg.

"Ah, Greg, don't forget that you have a luncheon meeting with David Freetman at 1:30. If you have any profit calculation or other financial facet, please take it with you, it may help. David is detail-oriented" said Tina.

"Thanks for the advice," replied Greg.

Greg arrived at Brennan's restaurant at 1:20 p.m. Valet parking took care of his car and the restaurant receptionist asked him if he had reservations. When he responded that he was supposed to meet David Freetman, she said, "I have a table for you. David called and wants to apologize for being about 10 minutes late." Greg sat down and ordered a Campari on the rocks. Since he received an email from AROBCO's purchasing department just before leaving the office, he took the opportunity of David's tardiness to answer the mail. He held his iPhone under the table so he would not disturb anybody in this fancy restaurant.

As David Freetman arrived he stood up and they shook hands. "Well" said David, "you and Vernon have similar tastes," and waved the waiter ordering a Martini. They sat down and David said, "I heard that you know Ray Villaloberg."

"Yes, I worked with Ray. This was several years ago," replied Greg. They were interrupted by the waiter and placed the order.

As they waited for their meals David continued, "I spoke with Mike Jacksens. This new control system order is not only promising from a future purchase order perspective but also very profitable."

"Yes, these types of systems can have a high profit margin, but they also require close application follow-thru. There is little room for error," responded Greg.

"Here come the meals. Bon appetite, Greg," said David.

"Bon appetite to you, too," rejoined Greg.

During the meal, David came to the point of the meeting and said, "Greg, I assume that you talked with Vernon yesterday."

"Yes, and that I was surprised by the turn of the events is an understatement," answered Greg.

"Yes, it is unfortunate, but we cannot dwell on this. We need to move on. Vernon and I had a long discussion and we decided that you should be the top person in command of MICGEN. That is, if you accept the offer," said David.

Greg decided to take control of the dynamic and said, "yes, I am delighted to take your offer. I do have some questions I would like to ask to get an understanding on the financial responsibilities. The hardware of such a project like the new control system order represents less than 10% of the total dollars. Our success and profit margin of such jobs is largely determined by timeliness and quality in both hardware and application. Besides, hardware delays and component quality have not only a drastic impact on overall profitability, but also on customer satisfaction. Thus, my recommendation is that Paul Bingam, who specifies the components and the primary manufacturer, should approve whether alternates are to be considered or not. This could change on a project-by-project basis. Would you agree with this?" asked Greg.

David Freetman's face was set into an expression of intense attention, like a man listening to an important broadcast which might affect his course of action in some way. After all, he always preached to Mike Jacksens and Al Murrell that they have to negotiate every component purchase contract to achieve the maximum discount. He responded to Greg, "I have not looked at it from the angle you just described. You know we financial guys are questioning any figures that have not been thoroughly negotiated. OK, let's go with your approval suggestion. By the way I do have a copy of the profit calculation details. Looks like you came prepared," said David, pointing at the papers Greg brought with him. Then he took a deep breath and continued, "alright, there is one more thing I wanted to talk with you about and that is the percentage of ownership. With the recent award of the large Safety System project and in view of the future AROBCO business potential, MICGEN's value has significantly increased. In consideration of your contribution, Vernon and I discussed sharing a percentage of ownership with you for a minimal investment. Presently the shares are divided 67% to 33%. Our proposal is to offer you 10% ownership for $50K. This would mean a 60 –30 -10 ownership division.

Did Vernon explain to you that we are a Limited Liability Corporation, an LLC? There is no stock; instead owners have a membership interest. The members, Vernon and I, and you if you choose to, do not have liability for the company's obligations unless you have signed personal guarantees. There is no need for that in this company. Profits are not taxed at the corporate level. Instead, members must pay taxes on these profits. I also want you to know that I am open to increase your membership interest in the future.

Think about it and let me know by the end of next week," said David. He carried on, "oh, and I almost forgot an important subject. Greg, there are several companies interested in MICGEN. I may come by the offices next week to show a company from the UK around and they may have a few questions, just so that you are prepared."

"Oh, no problem, as long as you advise how many people I can expect. This way I can tell the personnel that a potential customer is visiting," replied Greg.

"Of course I will," said David and looked at his watch. "I am pleased with our meeting, Greg. I look forward to working with you," said David. They stood up to shake hands.

"My pleasure and thanks for the lunch," replied Greg, and both left the Restaurant.

When he arrived back at the office the receptionist greeted him with a sociable, "hi, Greg, had a late lunch?"

He went right to Tina's office and asked her if she had time to go over Vernon's list in his office. As they went through the items, Greg remarked, "there are certainly many customer calls to follow up. Could you put a summary together of your valuation of each of these clients; just a paragraph or two on each: our status with them, their preferences, etc.? I know that's a lot of work but I think it would really help me to have some background of our customers before I make the contacts."

"Sure, actually it will not take me very long to make this summary," said Tina.

"Considering all this, I don't think that Vernon's idea of telling the employees and customers that he is taking a leave of absence is practical," Greg said.

"Yeah, you are right. Vernon was in a very emotional state. His wife told me that he intends to come in tomorrow night to clean out his desk. Maybe you could call him or meet with him and try to convince him to make a joint company announcement on Monday. However, from what his wife indicated, the chances of that are not good; he just wants to move on. He is starting at the new company in Raleigh, North Carolina the following Monday."

"Yeah, he will be the President of a company three times this size. I will see if I can reach him at home tomorrow," said Greg.

Then Greg asked Tina, "What do you believe the reaction will be to the proclamation of Vernon leaving and me taking over?"

"You have a lot going for you Greg. Your R&D team is motivated and they are letting everybody know about it. My advice is: do the same thing company-wide as you did for your R&D staff. Ask individual team members what they would like to see happen in the organization. Everybody wants to feel heard and understood," said Tina.

"Yeah I know, but when one has been here only a few months, like me, it is difficult to gauge the overall personnel situation," replied Greg.

"Largely, you will find that we have good circumstances in this company. But be aware of a few rotten eggs. Sometimes, despite your best efforts, you have a few people in this company who never want you to succeed; I have seen this with Vernon here. In your case it probably will extend a bit further; it could be that they wanted the position awarded to you for themselves," says Tina.

Greg recognized that this will probably be the case and comments, "I would address the issue directly with the person, and if he or she refuses to be cooperative, I may have to get rid of the person." Greg then looks at his watch and realizes that it is past 5:00 p.m.

He apologizes to Tina, "Oh, I am sorry, I held you over way too long. Again, thanks for the advice and have a good weekend."

Tina replied "You too, Greg".

Assurances after the president's resignation

Monday morning Greg let Tina know that he was not successful in convincing Vernon to make a joint declaration and asked her to get everybody together in the conference room for a short termination announcement at 9:00 a.m.

At 9:05 Greg walked into the meeting room, paused for a moment, and said, "Hi, everybody! Unfortunately Vernon could not make it, so I am here to tell you that he has resigned and I am taking his position. Vernon's termination was completely unexpected and it took me by surprise. And I would think that you are stunned, too. Here is what I can tell you:

I am not going to be the 'new broom' that sweeps any previous actions away. I want to assure you that it will be 'business as usual'. I will be talking individually talking with each of you later today about how we want to work together; what your expectations of me as your manager are; what hopes, fears and aspirations you have; what your motivators and de-motivators are, etc. And, I will also ask you what you think needs to be done to make the team or department more effective.

I want to encourage everybody to work together. The outputs of an effective team will always be greater than the individual outputs of the team members. I would also like you to know that I am learning the management game and that I will need your support as well. Thanks, and I look forward to talking with you individually."

Everybody looked startled and shocked, some of them even open-mouthed, and the whispering lasted for several minutes before they left the conference room. The situation was awkward and Greg left right away. He asked Tina to come with him to his office and told her, "I realize that Mike Jackson and John Kramo most likely did not appreciate me going over their heads directly to their staff, but that was my calculated risk. Besides Mike and John, who would you suggest I speak with first? Can you please take our telephone list and simply put a number next to each person. I am eager to talk with each individual, but I would like to get through with these 'pulse-taking' dialogs by the end of the day."

"Will do," Tina said. "Yes, and I would like to talk to Jim Berryson. How can I get a hold of him?" asked Greg.

"Jim is currently in Abu-Dhabi. There is a nine hour difference. His hotel number is 011 971 3-4001." replied Tina.

Greg did not want for Jim to find out about Vernon's resignation from somebody else, so he phoned right away. They talked for almost an hour going over Vernon's direct involvements with clients, customer follow up needs and other urgent damage control actions.

After he got off the phone with Jim, Greg spoke with Mike Jackson and John Kramo in their offices and then went from cubicle to cubicle. With the production staff he held the conversation individually in his office. At these separate consultations he offered a genuine smile to each employee, spoke briefly about Vernon – his move to a large company and his emotional state being the reason for not coming in today. He casually repeated the points he made at the 9 o'clock announcement - working together, expectations of him as manager, hopes, fears and aspirations, motivators, and cautiously probed each employee what he or she thought should be done to increase the team's effectiveness. He listened and took notes. He told them that certain talking points will be covered in this Friday's company meeting—the company's goals, values, and products. At the end he let them know that he is available to help and that he is looking forward to working with them.

In the afternoon Greg started the challenging effort of informing the company's customers of Vernon's departure. He studied Tina's description of each client - their personalities, 'hot-buttons', MICGEN's status with them, etc. He added notes based on Jim's input. He recognizes that when somebody like Vernon – a client-champion – leaves, it is almost like starting over with the customers. But Greg put the emphasis on the upside. It is his opportunity to reiterate his company's strengths and his commitment to quality service, as though they're a new account.

He phoned customer after customer. In some cases it took more than three attempts to reach the client. Greg explained to them that Vernon left the company to assume the presidency of a large company that is in a different business and reassured them by emphasizing that the company's commitment to responsiveness and overall client satisfaction has not changed. He also told them that he wants to be their main point of contact and comforted them that they're high priority. He also took this chance to ask, "What can we improve?", while setting up a time to have coffee, lunch or another meal with his clients. He stayed positive in re-establishing relations with his clients and re-emphasized the advantages of working with MICGEN.

At the end of the day, Greg was exhausted, but elated. He allocated a brief time for thanking Tina for her good advice. It will develop into a routine that includes corporate camaraderie and sets up the foundation for a successful tomorrow.

During the first week as head of the company, Greg spends significant amounts of time with each individual employee. He knows that it will take time to win their respect.

Friday morning Greg told Tina to call an all company meeting for 10:00 a.m. and to distribute the meeting agenda beforehand.

All Company Meeting Agenda

Date/Time: 10:00 a.m. Friday

Estimated Duration: One hour

As our company grows, the individual team leaders cannot do it all themselves. In fact, it becomes more and more important for all employees to understand the company's values, goals and strategy and for the people closest to the problems to take the initiative to solve them.

Doing that well means that we must maintain the same level of transparency when we have 35 employees as we did when we had only 7.

Values - With the recent contract awards we will need to hire five new people. It's crucial for these individuals to share our values of getting results, integrity, team-work, building inspired products, and putting customers first. But it's also critical that our teams act according to these set of shared values.

Goals - While values provide a valuable context for what people do, we must be grounded in the articulation of key goals: build a great product, grow consistently, delight customers, and achieve long-term customer services.

Strategy - Goals are important, but without a strategy, how will we get there? In order for each member of our team to take initiative, all of us must understand the strategy.

- Arenas - Greg will explain which products we plan to sell and which customer groups our firm will target.
- Vehicles – Greg will clarify why our growth will come mostly from internally-developed products & services.
- Value Proposition – Greg will describe how we are going to convince potential customers that the benefit of buying our product exceeds the price we charge, and the value propositions of our competitors.
- Economic Logic – Greg will make recommendations how we can increase our profit margins.

Initiatives – Highlights of progress being made on the new product development. Paul and Ken will discuss their progress, what they plan to do in the weeks ahead, and any challenges they are encountering.

Results - We will discuss results --sales versus targets, the number of new quotations, competitive wins and losses, number of new employees.

People are encouraged to ask questions about any of the topics!

A manager's typical days: The next four months of Greg's office life consisted of typical days. Actually, there is no "typical day" for the manager of a small company. Projects and products are multi-disciplinary organizational efforts involving various people inside and outside the company. Project and product life cycles call for different skills and people at different times. The issues and challenges start-off numerous and evolve throughout a project or product. It is difficult to characterize a typical day under these circumstances. Laced within and among other activities is a great deal of communication -- on the phone, via e-mail, in meetings, teleconferences, memos, and reports. And, of course the client is king, requiring ongoing attention. Small company management is a team sport and involves wearing multiple hats.

Over that four months, several developments took place at MICGEN:
- Mike Jacksens resigned.
- Two test technicians were hired.
- Jerry Fawvor, a senior control systems application engineer, was recruited.
- Ken had to change apartments twice due to his loud piano playing.
- John ended up in the hospital again, but is back at work now.
- Eight Requests for Quotations (RFQ's) for integrated Safety and Control Systems were received.

- Purchase Orders for four Safety Systems were received.
- Three Safety Systems were delivered.
- Three Advanced Control Wizard (ACW) prototypes were in test.
- The Alarm Management Software development had been completed.

The item really missing on the list was the Purchase Order for the AROBCO Production Platform Control and Safety System. A bid value of ~$8 Million. A project that would propel MICGEN into the big league.

The proposal due date for the AROBCO Production Platform Control and Safety System had been delayed for five months, luckily for MICGEN, because the new system was not ready until now. However, the sales strategy had become more complicated due to personnel changes at the Process Managers level. While Ray Villaloberg was still in the key position of VP Instrumentation and Electrical, the new Process Manager, Adam Morisen, assumed a very active role in the control system selection. Ray was still the actual decision maker who could approve the purchase, but it would be difficult for him to do so if Adam Morison had certain objections.

The AROBCO compression-separation unit Advanced Control Wizard (ACW) would be put in service in one month and the proposal for the total Production Platform Control and Safety System was now due in two weeks. Greg planned well for the presentation. After all, there was a lot at stake. With the new system prototype ready and a good PowerPoint presentation, he felt confident. He had analyzed what the customer wanted and focused on the needs of the customer, not the product which at this stage had no reference installation. He knows from Ray Villaloberg's feedback what is most important about the product to AROBCO and he would stress these values. He knows this needs to be an interactive, dialogue-type presentation.

'OK, Greg you are ready,' he said to himself. A little bit of nerves are okay. He is confident because he had put real thought and effort into this system presentation. He knows his product. He knows the buyer. He is ready to listen. He is solving a real problem, and he is ready for any objection.

Greg is walking back and forth in the conference room, looking at his Advanced Control Wizard (ACW), a device with triple modular redundant architecture. Each of the three systems were not much bigger than a modern mobile telephone, plugged into a mother board the size of the palm of a hand. Greg studied this object of so much effort. "This could change the future of control systems" he whispered to himself. His belief in the idea, combined with his persistence, had carried him where he is today, being the president of a company that is technologically several years ahead of the competition.

The Control Wizard module was located on the center of the conference table, next to two stainless steel enclosures, one being explosion-proof, housing the Intelligent Termination Panels. The user interface, a note-book PC was situated on the end of the table, near the large LED screen. On the screen

Greg had a slide showing the Esmix compression and separation process with a picture of the ACW in the center; it looked almost like a spider with its legs connected to various transmitters.

Greg was waiting for the AROBCO group to show up. Ken Beamer was sitting at the end of the table, checking out fuzzy-logic functions that he just added to the program, but they had nothing to do with Esmix. Tina was bringing in a pot of coffee, and as she looked out the window she saw the customers stepping out of their car in the parking lot. She alerted Greg, "they are coming."

Greg said, "OK, Ken, stop playing with the fuzzy logic and switch to the Esmix program."

"Done, I will be waiting outside the door for your call," replied Ken.

Tina went to the reception area, welcomed the AROBCO team and showed them to the conference room. There Greg greeted them with, "Good morning, everyone. Welcome to our presentation of the Advanced Control Wizard for your Esmix compression and separation process." He shook hands with everybody. Ray Villaloberg hugged him and introduced his team: Tom Deaverer, Vice President of Production; Adam Morisen, Process Manager; and Roul Garciabo, Maintenance Manager.

Before they all sat down, Tom remarked, "well, that looks familiar", pointing at the process on the LED display. Greg is trying to size up as much as possible, their mood and energy and some sense of the relationships between Ray and the others. He then talks about some general commerce MICGEN had with AROBCO and then moved to the main item, their compression control system.

He started off by saying, "First, let me address the benefits of the device as related to your process application; the main being process disturbance avoidance. During low throughput operation, a gas well change can cause significant fluctuations in separator pressure, flaring, and in some cases even a process unit shutdown occurs. The Control Wizard will practically eliminate these process upsets.

Secondly, during throughput of less than 78%, presently, the compressor recycle valves start to open. The Wizard will allow you to operate below 60% before the recycle valves start opening, saving a significant amount of energy.

Thirdly, on high throughput situations the separator limit pressure will be automatically adapted to maximize throughput, resulting in increased plant capacity." He added, "we have prepared a simulation that I believe will allow us to show you how these issues are addressed in detail. If you have any questions, please do not hesitate to interrupt. We are here to answer your questions and address your potential concerns. That is what I see this meeting is about."

"I have asked Dr. Ken Beamer to give us a demonstration of the Advanced Control Wizard (ACW) with a simulated process," Greg said. "Ken has been leading the ACW software development and he is also a chemical engineer." Greg then opened the conference room door and called Ken.

Ken introduced himself to the management team and went through the process using an electronic pointer and the PowerPoint process slide. Tom Deaverer and Adam Morison leaned forward, indicating a certain enthusiasm for Ken's understanding of the process. Ken then changed to the second slide and presented the simulation process showing the process model, including curves, response times and the I/O. He also gave an overview of the ACW program which executes input validation, steady state target calculations and dynamic move calculations, and has multiple model identification algorithms plus model prediction, model uncertainty and cross-correlation features for model analysis. When Ken saw everybody staring at the ACW he said, "we will go through the additional controller features later, if this is OK with you." They all nodded. Then Ken carried on, "our experience has shown that today's control systems cannot handle this unexpected process behavior in a reliable and safe manner for the following reasons:"

- Typical DCS systems do not have the required fall back function in case of certain transmitter malfunction.
- These systems also lack the adaptive tuning capability.
- They do not have constraint limit calculations.

"Are you saying that in the event of a transmitter or analyzer failure your system compensates for it?" asked Adam Morison.

"Yes, it does automatically switch to a fallback algorithm. Let me give you an example: Let's assume that on your anti-surge control you have a pressure or temperature transmitter malfunction; it will automatically transfer to minimum flow control. Let me show you that on the simulator, if you don't mind?" said Ken.

"Go ahead," replied Adam.

"See how this transferred. And here is the alert confirming it to the Operator. OK, while we are at it, let me show you the effect of the analyzer failure on the Separator. See, in both cases the control continued. While the fallback strategies may not be as efficient as the primary strategy, the main thing is that there is no interruption of the process in case of certain instrument malfunctions," said Ken.

"Yeah, this feature is, from a reliability standpoint, almost as important as the triple modular redundancy," added Roul Garciabo.

Then Ken continued, "there are several other key features, such as Constraint Limits, which are from a safety and reliability standpoint significant. If you like, I can present this to you on the simulator," said Ken.

"It's OK Ken, we know now that you can relate to both your system and our process," remarked Tom.

"Thanks, gentlemen. Greg do you want me to show any other functions?" asked Ken.

"No, Ken, but we should perhaps cover the communication interfaces," said Greg.

"Yes, I have several questions regarding interfaces," said Roul. "Can you accommodate intelligent transmitters?"

"Yes, as a standard feature, the ACW handles intelligent and HART transmitters," replied Ken.

"How about your communication to the operator interface?" asked Roul.

"It is H2 Ethernet, Intranet, Internet and there is even a Bluetooth interface. These communications are standard. So, from a maintenance perspective you have remote monitoring built in. We also have a Firewall security chip that keeps the interfaces safe from viruses."

"Thanks for that comprehensive reply Ken," said Roul.

"Would you gentlemen like to have additional information?" asked Ken.

"No, thank you Ken for your excellent demonstration," said Ray.

Greg stood up and said, "thanks Ken. I would like to just briefly summarize the benefits of our product, if you would allow me to do so."

"Go ahead," said Tom.

Greg then covered the advantages of the system as they applied to AROBCO's process and highlighted that these benefits would not only be applicable to the compression separation unit, but also to the whole production platform process if AROBCO accepts MICGEN's proposal.

Ray then said, "Greg, we thank you for your outstanding presentation. Unfortunately, we have to leave sooner than anticipated because we have to make an unplanned stop. Again, we appreciate all your good work."

"You are very welcome and have a good return trip, gentlemen," Greg said. They shook hands and left.

Greg went to Ken's cubicle and said, "outstanding job, Ken."

"Hey, from all the time we spent going through that process and these fallback functions, something had to rub-off," replied Ken.

"I could see by watching the AROBCO team's body language that they were very pleased by your performance. Don't undersell yourself," said Greg.

Greg wanted to discuss a few specific functions with Ken, but was interrupted by the paging announcement "Greg, Ray Villaloberg on the line for you." Greg ran back to his office and picked up the phone saying, "hello Ray."

"Hey Greg, we forgot to talk about the enclosure types. Sorry we had to leave early. I am here in the car with the guys, and we still have more than a half hour ahead of us, so perhaps we can sort a few more things out if you have time. Regarding the enclosures, is the ACW intrinsically safe?" asked Ray.

"We have applied for certification at CSA and ATEX but have not yet received the approval confirmation. The Intelligent Termination Panels have been, however, certified," responded Greg.

"Yes, we know that the IPT's are certified I-Safe; we have them in our plants. When do you expect to receive the I-Safe certification for the ACW module?" asked Ray.

"From CSA within a few months, from ATEX I am not sure," replied Greg. Greg heard them talking in the background, but he could not understand the discussion; then Ray came on the phone again, saying, "we already have your Ex-Proof type enclosures and decided to stay with them for the compression/separation unit as well as for Esmix plant-wide. We will change the production platform RFQ to reflect this and send it out to everybody in the next couple of days. Thanks, Greg. That was it."

"OK, was everybody satisfied that they got their questions answered at the presentation?" asked Greg.

"Yes, hey, hold on to your whiz-kid, Ken. He can answer anything! If there are additional questions I'll let you know right away," answered Ray.

"Again, you guys have a good trip back to the UK," said Greg and he hung up.

The customer's familiarity with the ITP (Intelligent Termination Panel) was an item Greg overlooked in factoring the decision of the AROBCO Esmix quotation. Oh, and there were also customer testimonials on the high-reliability of the ITP, which Monica showed him several months ago. 'How could he have not included all this in his proposal on the Esmix quotation' said Greg to himself. Yes, he included the IPT's in the bid, but the fact that hundreds of these have now been in service for more than a year without any malfunctions needed to be highlighted. Also the high temperature range of up to 70 degrees Celsius, due to the use of military spec components, should be stressed. In addition, the fully redundant high-speed serial I/O link must be emphasized. Greg knew from experience that these types of details can make the difference of a successful or unsuccessful proposal, so he called up the quotation folder on his PC to make the changes.

As a result of meeting AROBCO's technical requirements, exceptional customer relations and attention to details, Greg's proposal was successful. MICGEN received the huge purchase order, but did not have the money to fulfill it. Greg needed to hire workers, buy raw materials and meet production expenses. Of course, he knows about all this and he is also aware that the order exceeds David Freetman's, the principal owner of MICGEN, current financial capabilities and that the bank won't approve a loan. While getting the large purchase order was rewarding, it was also going to be very challenging.

To minimize this financial problem, Greg, with David Freetman's consent, applied for purchase order financing for the hardware portion of the project at the time he was quoting the job. Purchase order financing is designed to help companies that have growing purchase orders but don't have the financial resources to fulfill them. It provides funding to pay suppliers and enables a company to accept and deliver large purchase orders. However, purchase order funding is not for everybody. To qualify for funding, your company and your customer must have good commercial credit. This was the case with MICGEN and AROBCO. Although this type of financing normally applies to companies that resell finished goods that have been purchased from a third-party supplier, there are some exceptions. Additionally, the transaction has to meet the following criteria:

- It must have gross profit margins of at least 20%.

- Your supplier must be capable of fulfilling your order.

Purchase order funding transactions are structured as follows:
- Your company gets a large purchase order from a credit worthy customer.
- The transaction is submitted for review and approval.
- Your supplier is paid, usually through a letter of credit or similar instrument.
- The goods are transported and delivered to your customer.
- Your customer pays the invoice, at which time the transaction is settled.

Of course, Greg accepted the Purchase Order, a project he pursued so intensely. This purchase order constituted his offer to AROBCO and now becomes a binding contract upon the terms and conditions stated in this purchase order.

Although he was not 100% sure, he acknowledges that he has in his possession all applicable specifications, drawings and documents necessary to perform the obligations under this purchase order at the price and schedule stated on this purchase. And Greg was fully aware that this purchase order was not secure if the startup of the Advanced Control Wizard prototype at the Esmix compression/separation unit was not successful.

Like so many small businesses, MICGEN eventually faced the issue of handling business expansion. Business expansion is a stage of a company's life that is fraught with both opportunities and perils. On the one hand, business growth often carries with it a corresponding increase in financial fortunes for owners and employees alike. In addition, expansion is usually seen as a validation of the entrepreneur's initial business startup idea, and of his subsequent efforts to bring that vision to fruition. But as Greg experienced, business expansion also presents the small business with a myriad of issues that have to be addressed. Growth causes a variety of changes, all of which present different managerial, legal, and financial challenges. Growth means that the company's management will become less and less centralized, and this may raise the levels of internal politics, protectionism, and dissension over what goals and projects the company should pursue. Growth means that market share will expand, calling for new strategies for dealing with larger competitors.

Growth also means that additional capital will be required, creating new responsibilities to the Entrepreneur/Manager and his principal investors. Thus, growth brings with it a variety of changes in the company's structure, needs, and objectives. Being relatively new in management, it took Greg some time to accept this reality.

The Onsite Startup

With all the challenges he had overcame in developing the new product and in getting the purchase order financing for the big order, Greg knew that the upcoming Advanced Control Wizard (ACW) trial installation at the Esmix compression/separation unit can make or break MICGEN's success show. A failure at Esmix would certainly cause a cancellation of the big order. It would be devastating for MICGEN. Thus, as far as Greg was concerned, this trial was the final hurdle to success. A fireworks display is only spectacular if the final moments are the most awe-filled of the entire show. In the world of control system implementations, the big-bang ending, ironically, is startup. But a great startup doesn't only happen in the final hours. It's a culmination of carefully orchestrated activities throughout the development of the control system and the project.

Greg knew from experience that planning is a key aspect of a successful startup. He made certain that all facets of the project had been carefully planned and documented, including all training and testing protocols, and that the dynamic simulation was detailed and thoroughly exercised. This simulation effort was vital to cutting the site implementation time drastically, and both Greg and the client, Ray Villaloberg, were counting on this. The simulation positioned Ken and Jerry, the MICGEN startup team, for a successful transition from the old system to the ACW (the Advanced Control Wizard). Greg had confidence in his team. With Jerry Fawvor, an experienced field engineer, at Ken's side, he knew he had a competent crew on site.

Greg also realized that thorough pre-verification before the field startup was crucial and he scheduled for it. For instance, Ken would be stationed at the human-machine interface (HMI) in the control room while Jerry was at the field instruments. In addition, the team included an electrician from the customer. The group moved through the compression-separation part of the plant ensuring that all devices checked out properly from HMI animation to the correct device in the field starting, moving or rotating as expected.

Greg's, Ken's and Jerry's collective knowledge of the client's process was another critical aspect of startup success. Ray Villaloberg, the VP of the client, commented, "the one thing I appreciate most is that the MICGEN team has a deep understanding of our process. When the supplier understands our process, I don't have to worry about an accident happening." Industry knowledge and onsite support has even more advantages for the client when the startup is for a retrofit, like in this case. In-depth support for a retrofit project is so important since you never know what undocumented features may be uncovered as you change over to the new system.

The atmosphere in the control room was tense, as it usually was during startup operations. Operations ran through the check list. The compressor started on full recycle with its status monitored.

"What is the Gas Well status?" asked Joe, the supervisor.

"Ready to go," replied the field operator. The ready to start light glowed green. Gary, the senior operator, said "OK, everybody ready?"

Jerry Fawvor, with his over ten years of field experience, knew this was the make or break moment for the Control

Wizard. The perfect outcome for these first minutes was to see and hear nothing, no yellow or red lights on the emergency annunciator panel, no horn. His panic scenario was hearing the relief valves go off. Ken, on the other hand, appeared sure-footed in the surroundings. To him, it was like a repeat of the simulation, the seventh 'imitation'. Sometimes, inexperience has advantages. He was having fun and his anticipation grew as he imagined what might come next, since his simulations had many states.

As the production platform restarted, the plant operators were nervously looking at the separator pressure, expecting the big fluctuations that usually occurred during the startup and well switching phase. One operator was standing at the compressor ESD (Emergency Shutdown) button ready to stop the compressors in order to prevent overpressure flaring. The separator pressure increased at a rapid rate and then flattened out. No fluctuations, no compressor trips. The usual process upsets during this critical operation phase did not happen. The anti-process disturbance rejection algorithm of the Control Wizard worked. Jerry Fawvor, usually reserved in his behavior, showed his excitement with an elbow poke into Ken's ribs. The operators and the shift supervisor had a hard time believing the smooth process behavior, considering the typical instabilities during every one of their previous startups.

Joe, the shift supervisor, could not contain himself. "Wow!" he shouted and jumped up from his chair. "Super! There is a first time for everything. What a great startup." He then set back in his chair and smiled, thrilled at the thought of having such hassle-free startup operations in the future. He waved at Ken to come over and said, "hey whiz-kid, how did you do it? How come you don't even seem to be excited?"

Ken replied, "I did not do anything. The Control Wizard did it. No worries, per simulation it had to behave."

Joe shook his head and repeated, "it had to behave per simulation."

Jerry stepped in to rescue Ken, saying to Joe, "Ken meant to say, our effort to thoroughly check out and simulate the control system software and hardware in-house, to the maximum extent possible, before going to the field for the actual startup has paid off."

Success

At 5:00 a.m. Greg received a phone call from Ray, "hello Greg. Hope I did not wake you up. I know you are an early riser, but it is only 11:00 a.m. here, 5:00 a.m. your time."

"No, Ray, I was already up waiting for a phone call from my troops. Things went well at Esmix?" asked Greg.

"Things went very well. The startup was smooth and they are presently running at 117% of capacity. Operations induced throughput changes to verify the process and equipment behavior at different conditions. They were running at 65% with the recycle valves closed. Would you believe that? And now they are at 117% with the separator pressure at auto constraint, and it looks like they may be able to go to 120%. Fantastic!" said Ray. He added, "CAISTOS is about to kick off the implementation phase of their platform control modernization and you can expect to receive a phone call from your old friend, Hank Sandover. As you know, their requirements are much larger than ours here."

"Well, that is great news, Ray. Thanks," said Greg.

"Yes, it really is. I believe that you will have worldwide opportunities on all gas production platforms. I intend to contact the editor of Gas and Oil Magazine. He has been bugging me about new developments at our company anyway, but I will wait a few days. You should see something in their on-line issue soon. Have a good day Greg," said Ray and he hung up.

A week later the following notice appeared in the Gas and Oil Magazine. "The Advanced Control Wizard system of MICGEN provided the leverage needed to improve yields, increase throughput, reduce energy usage and achieve more uniform product quality. Optimal control and safety are critical requirements for all industrial plants, particularly in the oil, gas, chemical, and power industries." said Ray Villaloberg, Vice President of AROBCO.

Shortly afterwards, Greg received a call from Ray – "Hello Greg, have you seen my statement in the Gas and Oil Magazine?"

"Yes, I sure did. Thanks for the great endorsement Ray," Greg responded.

"Yeah, now they are asking me to provide an article on the benefits of Advanced Process Control; I just sent you an e-mail to show you what I prepared, please give me your comments as soon as possible," said Ray.

Greg checks his mail and opens Ray's message to find the following:

Advanced Process Control and Real-Time Optimization can benefit your Plant
By Ray Villaloberg, Vice President of AROBCO

Whether your operation has been running for 10 years or is about to start up, you face the ongoing challenge to achieve the highest returns possible on that investment. Once you have the primary bugs worked out and achieved the goals of targeted productivity and quality, how do you improve performance? Advanced Process Control (APC) has gained increasing attention over the past few years. It is key for success in increasing process stability and yield, while minimizing costs.

Usually, Advanced Process Controls and Real Time Optimization implementation take place during the steady-state phase of a process unit's life-cycle. Plants are however increasingly looking to expedite the benefits of Advanced Process Controls implementation by introducing it at the early stage and having it functional within months after the plant becomes operational.

Process industries utilize a vast and interconnected set of technologies and processes. The key challenge for refining, gas plants, chemical and petrochemical plants, etc. is to maintain processes at their optimal operating point while simultaneously maintaining multiple safety margins at acceptable levels.

AROBCO recently implemented a model-based feed-forward multivariable control system, provided by MICGEN, that monitors many process variables simultaneously and in real-time. Its Advanced Process Controls and Real-Time Optimization scheduling and management functions also offer a proactive view of plant performance; allowing operators to push production limits without jeopardizing equipment limitations.

Without any modification of physical plant hardware, MICGEN's Advanced Process Control offers both tangible and intangible benefits.

Benefits of Advanced Process Controls (APC):

- Improved production by securely reducing the safety buffers required to ensure that limits for product quality and equipment integrity are not breached.

- Minimized energy consumption for maximum plant throughput

- Stabilized plant operation through minimized instability of key process variables

- Improved responsiveness to changing economic and regulatory conditions through easy review and modification of operating goals

- Less unpredictability in the feedstock to downstream units

- Improved operator effectiveness by concentrating attention on the key unit performance indicators

- Enhanced process safety as the APC system acts as an early-warning system

- Better understanding of complete unit operation

MICGEN has broad Advanced Process Controls experience and has implemented its Advanced Control Wizard (ACW) system in our plant on time and on budget.

Advanced process control offers substantial improvements to industry, but widespread misunderstandings and a lack of knowledge have hindered its implementation. The achievement of APC has also often been limited because most of the process control systems (DCS, PCS, etc.) on which the APC resides do not have the proper safeguard and automatic fallback strategies that provide for loop integrity and reliability.

Real-Time Constraint Limit Control (CLC) and Optimization:

The real-time CLC, provided by MICGEN, is a complex, rigorous model-based system that complements the APC installment and improves performance by adjusting a series of process variables in order to increase profitability and minimize operating costs.

It contains:

- The Histogram and Normalcy Probability Plots for continued Normality Test

- The Measurement Discrimination Evaluation

- The What-If Analysis Routines

- The Constraint Soft and Hard Limit Value Calculations or Pre-Settings

- The Process Unit Efficiency Calculation Program

- The Automatic Control Fall-Back Strategy Selection

Overall Benefits of an Efficient Process Control System:

Your plant can benefit from a well-designed control system in many ways, including…

- **Energy Savings** - energy wastage is reduced when your plant and machinery are efficiently operated

- **Improved Safety** – advanced control systems automatically warn you of any abnormalities which minimizes the risk of accidents

- **Consistent Product Quality** - variations in product quality are kept to a minimum and reduce your waste

- **Improved Environmental Performanc**e - systems can give you early warning of a rise in emissions

Achieving the business goal of maximizing profits and overall value requires a balance between many factors. The time is right to implement process optimization. In particular, Advanced Process Control improves operating stability and constraint handling resulting in less unscheduled plant downtime, generating more throughput, saving substantial plant and energy costs and improving profitability.

End of Article

Greg could not have asked for a better commendation from Ray and immediately called him, "Hello Ray, this is a great article. I want you to know how much we enjoy serving your needs and that we consider you a special customer. Of course we appreciate your orders, but we also appreciate working with you".

"Hey Greg, I have recommended your company to others because of our satisfaction with your product and service. I look forward to doing business with you for years to come," said Ray. "Thanks so much," responded Greg and they hung up.

The effective introduction of the new control and safety system technology and the successful startup at a customer's plant had been of great importance. But Greg recognized, once a product is launched, the key to profitable growth is to seek new sources of competitive advantage relentlessly.

In our rapidly changing world, competitive advantage is at best temporary and must be constantly pursued and renewed.

Funding

Now that the large control and safety system order was finally secure, there was no more cancellation danger. This would certainly help in obtaining more funding,' Greg reckoned. While he had been able to get purchase order financing for the hardware portion of the large project, the additional manpower requirements to properly execute the project were not adequately underwritten.

David Freetman's concept of operating on a shoe-string was not only causing him stress, but was also delaying a few small projects due to not being able to expedite certain component deliveries. Vendors do not want to assign high priorities to late paying clients. And, during the past weeks, Greg had learned the hard way that banks do not want to deal with the unfamiliar world of lending to technology upstarts. Bankers, a conservative lot by nature, shy away from risk. When there is enough money involved, they may be willing to bend and lend, but in the case of MICGEN, where the loan amount would be about $400K, they were not interested.

Thus, before he talked to out-of-state finance players that focus strictly on tech, Greg considered digging into his retirement fund to assist in financing MICGEN. He believes that tapping into his 401(k) to help on the company financial issues would be a basis to demand an increase of ownership. It was time for him to take the next step on the ownership issue and make the call, though cautiously, to David Freetman. If the exchange did not play out, he would simply end the conversation. Trying to sound casual and confident, he called and said, "hello, this is Greg. I am following up on our conversation regarding ownership increase, if you remember during our business luncheon."

"Oh yes, I recall. I can tell you, though, Greg, I am flexible but I am not going below 51 percent or in other words, I am willing to give up 9 percent of my ownership, but not more than that. You can talk with Vernon. Perhaps he wants to share a percentage of his ownership," said David.

"Thanks, David. I will try to get a hold of Vernon," said Greg. From the few words, Greg got the message as to how sensitive MICGEN ownership is. David Freetman did not even ask how much he was willing to invest or what percentage he was asking for.

Greg felt uncomfortable about the conversation with David. And as it is often the case when he wants to lighten the load on his shoulders, he seeks advice from his reliable friend Linda. 'Since she had ownership of a company for many years, she surely would know the subject well,' said Greg to himself and he called Linda. "Hi, Linda, sorry to bother you again."

"You are no bother. How are you doing? Have not heard from you in a while," said Linda.

"I am doing very well. The company is growing and I am thinking about increasing my minority ownership in the business," said Greg.

"I did not know that you are a part owner of MICGEN," said Linda. "What is you ownership percentage, if I may ask?"

"I have accumulated a 10 percent stake here, but I want to increase it to 25 percent. The principle of the firm holds a 60 percent stake in the company. Vernon owns 30 percent."

Linda immediately responded, "Watch out Greg! You would still have only a minority stake in a company where one person owns a majority interest, unless the principle owner is willing to give up more than 10 percent of his ownership. Do you understand the consequences of such a distribution of ownership?" asked Linda. "You have very limited rights as a minority shareholder, assuming that there is no written shareholder agreement addressing these issues. In the absence of an agreement to the contrary, you basically have the following rights as a minority owner: a) If the company is sold or dissolved, you get your proportionate share of the proceeds remaining after all debts are paid; b) If there is a distribution of profits, you are entitled to a share of the distribution; c) You have the right to demand an accounting, which is a limited right to examine the books and financial records of the company; and d) you have the right to sue for breach of fiduciary duty if the majority owner engages in bad misconduct.

I will not address the last situation, involving issues tantamount to fraud. If you believe the majority owner is defrauding you, see a lawyer immediately, and be prepared to spend some real money."

"Hmm, so what issues do minority share owners, like me, confront?" asked Greg.

"Well, here are the consequences of your bundle of rights as a minority owner when addressing day-to-day business issues," said Linda.

She continued, "as a practical matter, your right to demand current distributions from an operating business is limited. A majority owner, if he is committed to avoiding any distributions to a minority owner, can usually avoid making any distributions of profits by establishing generous reserves for future expenses, paying a salary to himself or his relatives at the high range of what is reasonable, pre-paying expenses, investing in new business or new equipment, or leasing expensive cars, etc. A majority owner can spend enough that there are rarely any profits to be distributed. So long as the expenses are not grossly unreasonable, you probably won't be able to force the company to allow you to share in any of the current income of the company. You have no right to participate in any management decisions of the company. The majority owner may make a decision that you think is bad and puts your interest in the company at risk. You may see the majority owner running the company into the ground. You can try to convince him that it is the wrong decision, but he doesn't have to take your calls," and Linda carried on.

"You have limited rights, if any, to have your interest bought out. You may want to cash out your interest and do other things with the money. State law may give you the right to force the company to buy you out, but these rights are limited. And while you would be entitled to share any profits on sale of the entire business, a sale can be structured in a way to avoid any payout to minority owners, such as a sale of assets over time with the proceeds reinvested in another business."

"So, how can I protect myself as a minority owner?" Greg asked.

"I am not sure that you can. The law of corporations and limited liability companies gives majority owners almost unlimited discretion in deciding how a company will be run and what rights any of the owners may have.

To address your scenarios, as an investor you may want to contribute only $20K in equity for your additional 15 percent interest and have the other $80K be a loan secured by the assets of the business and a personal guarantee of the majority owner. As an employee, in case you leave the company, you would be in better shape if the agreement provided for some mandatory buyout. You would, however, have to expect to take a sharp discount from the proportionate value of the interest in the company," answered Linda.

She continued, "the provisions regarding management of a company and protection of minority owners are limited only by the creativity of the owners and a complete discussion of the possibilities is beyond the scope of our conversation. When considering a minority stake in a company, think carefully about your expectations regarding the following: involvement in day-to-day management; involvement in decisions about fundamental corporate changes such as sale of the company; payments on your equity from current operations; when you will be able to sell or be bought out and distributions on sale or dissolution of the company.

Discuss these with David and Vernon, your business partners, to make sure everyone shares basically the same expectations. Then, you should go to a lawyer to have this understanding reduced to writing. While going to a lawyer to get an agreement will cost some money, it will be much less money than litigating problems once they arise."

Linda continued, "For your own good, Greg, let me tell you to think about all this and evaluate your aspiration for the 25 percent ownership carefully.

Sometimes, a minority interest is worthwhile and valuable, particularly when a business is sold.

A minority stake certainly does not mean that you inevitably will be shut out from any financial gain. If, however, the majority owner is not cooperative and no written shareholder or operating agreement exists, your minority interest in the company may not be worth much in any practical sense. In the absence of an agreement, you should realize that your ability to financially benefit from a minority interest depends to a great degree on the good faith of the majority owner. Thus, before you invest time and money in a minority stake in a business, think about whether trusting the majority owner is enough, or you want to get your agreement in writing to make sure that everyone's expectations are fulfilled."

"I don't know how to thank you for your valuable advice," said Greg. And Linda ended the conversation with her usual, "That's what friends are for," and hung up.

Dealing with banks

While Greg understood Linda's comments and decided not to pursue any ownership issues, he knew that access to capital is essential to his company. He recognizes now that rapid growth can hinder a company's ability to meet borrowing requirements. He needed a bank that could provide MICGEN access to additional working capital now to bridge his money necessity until shipment of the large project, when the partial payment from AROBCO is due.

Being aware that David Freetman's financial capabilities were not adequate and that David would not change the proprietorship sharing, he decides to call him and ask him to obtain a bridge loan from a bank. David picks up the phone on the first ring saying, "hello, Greg, what can I do for you?"

"Sorry to bother you once more on financial and ownership matters. MICGEN needs about $500K in funds. As you know the new order value is over $8 million. Despite the favorable payment terms and the purchase order funding for the hardware portion, we are unable to fund the costs of personnel and other items from what we have in the bank at this time. I assume that you do not want me to pursue any venture capitalist avenue and considering that the local banks have not been willing to extend a loan, what do you suggest?" asked Greg.

"Well, in this State, our chance of getting a loan is very low. We should try a bank in California or Illinois. Those banks increasingly view tech as a way to grow their loans without having to take customers from competing lenders. For tech startups like our company, the new attention from banks is a good thing, giving us more financing alternatives. I will call the bank in Chicago. You or I, or both of us may have to go there. I will get back with you in the next few days," said David.

Applying for a bank loan can be a frustrating and mystifying experience for small-business owners and managers. And finding the money to finance a growing company is one of the most difficult obstacles, as Greg experienced when he pursued the purchase order loan for the hardware of the new project. David was at home in the financial world and knew that it was key to zero in on appropriate targets. He looked for a bank that was familiar with his loan type, with the industry and one that had done business with people like him or companies like his. He sought out a bank that gave technology loans of the size and type MICGEN needed.

David called ahead to find out the name of the bank's small business specialist, and set up an appointment to meet in person. He asked the banker for a description of the materials he would want to review. He knew that typically, these will include a cover letter, completed loan application, business and personal tax returns, financial statements, and projections. In addition to the financial documentation, he needed to bring along an executive summary that detailed what MICGEN would use the money for and how they planned to pay it back. He also brought with him promotional materials about MICGEN's business - brochures, ads, articles and press releases.

Before he met with the banker, David prepared a three minute company PowerPoint presentation. He was familiar with the banking environment and anticipates their questions. He was confident and thoroughly prepared. He knew, of course, that the banker would ask him how much money he needed and how long he needs it for. He was prepared to go into detail about what he would do with the money and why he and his business are low-risk as well as when and how he would repay the loan. He would convince the banker of the long-term MICGEN profitability and his ability to repay the loan. David presented himself as a financial entrepreneur who could, and would, repay the loan. He kept it all real. He avoided broad, unsubstantiated statements in his loan application and kept projections, assets lists and collateral statements on the conservative side. He also discussed the risks with the banker to ensure that the banker knew that he thought about it and that he has planned for risks and intends to manage worst case scenarios.

David did not push it. When the meeting wrapped up, he asked the loan officer when he could expect the bank to make a decision, but he did not push too much. He knew that doing so might result in a rejection. He was aware that all he could do to ensure a speedy decision was to make sure that his application was complete.

MICGEN received a short-term loan of $700K that had to be repaid from the proceeds of AROBCO's payments - 80% 90 days after system shipment from factory and 20% at completion of deliverables. The term payments included principal and interest. David Freetman's bank negotiations were very effective and Greg congratulated him and thanked him for his efforts. The new project was well-funded.

End results

Many business discussions and meetings revolve and end up around money. Most of the engineers focus on developing and improving products. At first, these two agendas may appear to be at opposite ends of the spectrum, but they are not. Finding the common ground will enhance your ability to succeed in your business.

To stay competitive, machines and facilities have to be continually adapted to the latest requirements.

If the automation system is no longer up to date, it is time for a modernization.

In this process, each modernization has its own challenges. What is the individual goal of the client: faster return on investment (ROI), lower total cost of ownership (TCO), higher availability, or shorter downtimes?

No matter what the starting point of your customer is – there is often a generational change in the automation required. The modernization of systems, or the modernization of a complete plant through an Automation System can help your clients to reach their individual goals.

The end benefits for the customer at a glance:
- Higher productivity, total efficiency and usability
- Latest manufacturing standards, machine safety and industrial security requirements
- Minimized downtimes
- Increased profitability
- Improved competitiveness

Closing Comments:

This story relates to a small company involved in safety-related industrial automation solutions.

Today's economic pressures have affected performance and growth in industries utilizing process automation. Uncertainties regarding the current stagnation, fluctuating oil prices, globalization, and political forces are limiting manufacturers' ability to invest in new plants. Limited investment capital means that they can only undertake projects with short potential payback periods. Process automation offers the greatest potential leverage for improving productivity and profits. When properly designed and engineered, process automation solutions provide the opportunity to increase production rates, improve yields and reduce energy consumption.

However, automation works only in combination with human expertise. Process experience and application expertise are required to optimize the investment in a process-automation system. One must focus on the critical units in process plants (the key assets – industrial boilers, rotating machinery, columns, TMC for LNG, etc) where optimization of efficiency, reliability and safety can maximize the investment and return on capital expenditures.

While automation systems are often provided by major companies, these firms frequently do not possess the required experience for niche applications. Small and medium-size companies can deliver niche process automation solutions for end-users across a range of industries worldwide. They can leverage their systems capabilities and their application expertise and therefore offer executable solutions for many process challenges.

With the economy undergoing technological shifts, many industrial companies are finding automation preferable to hiring new personnel. A survey of Harvard Business School alumni released in September 2014 found that nearly half of the companies would rather invest in technology than hire or retain workers. While this displacement can undermine the wages and employment for low-skilled workers, it will increase the demand for engineers and scientists, which are high salary jobs.

The technological shifts will not only raise the demand for engineers and scientists; the focus on automation will also expand opportunities for technology entrepreneurs. The key is to select profitable niche markets.

And this concludes the story about "The Wizard of Automation". The purpose of writing the business-story section of this book was to create a cast of believable characters who were struggling to overcome typical problems in a technology company environment. The way they dealt with these problems will hopefully provide the reader with insights into the human condition in an entrepreneurial business.

Some readers may consider this book too technical and detail oriented. However, startups and tech companies that are carefully crafting technical and emotional experiences are the ones that are leading the way.

The goal of this book is to encourage entrepreneurs to strive for excellence and highlight the importance of details and application expertise.

Inference

Developing a new product is typically a multifaceted process. It can be thrilling and satisfying. But, with many technology products, there are huge challenges in bringing a product, no matter how genius the invention is, to market.

As a small company grows into a larger, established company, it faces the same pressures that make it necessary for today's businesses to find new ways to invest in innovation. In fact, an advantage of a successful small firm's rapid growth is that the company can keep its entrepreneurial DNA even as it matures. Today's technology companies must learn to master a portfolio of sustainable innovations. It is an outdated view that sees startups as going through discrete phases that leave earlier kinds of work— such as innovation— behind. Rather, a company must excel at doing multiple kinds of work in parallel.

Persevere - Over time, a team that is learning its way toward a sustainable business will see the baseline numbers rise and converge to something like what they once established in their business plan.

Sustainable growth is almost always characterized by one simple rule – Most new customers come from the actions of past customers. There are ways past customers drive sustainable growth:

- **Word of Mouth** - Embedded in most products is a natural level of growth that is caused by satisfied customers' enthusiasm for the product.
- **Effective Marketing & Sales Technique** - While there are no "one-size-fits-all" solutions to implementing an effective method, there is a key step that can be taken to help the company meet its goals: Gather customer testimonials and use them in the sales/marketing literature, in the advertising campaign and on the company website. When using these customer endorsements, tailor them to differentiate your business from the competition, also, emphasize benefits instead of features.
- **Through Repeat Purchase** – Pursue maintenance contracts and system upgrades.

Innovate - Conventional wisdom holds that when companies become larger, they inevitably lose the capacity for innovation and creativity. This is wrong. As startups grow, entrepreneurs can build organizations that learn how to balance the needs of existing customers with the challenges of finding new customers to serve, managing existing lines of business, and exploring new business models—all at the same time.

Hire the right people – This is most important for all firms. And when you are running a startup or a small company, the stakes are even higher. New employees will help shape your business in ways they may not be able to at a larger company. The hiring process may focus on the individual candidates and what they can bring to the table, but you need to think about how they will work with your existing staff, too. Hire the ones who are most passionate about your product or service. **People make a company; they are the company.**

And as a final perspective – As with any grand vision toward a sustainable technology business, the devil is always in the details.

Conclusion

Finally, entrepreneurship is neither easy nor risk free. We have all heard the statistics of startups failures. For entrepreneurs, this means that if you want to have a chance at success, you have to take risks. While risk is an integral part of entrepreneurship, it does not have to get the better of you. Great entrepreneurs achieve success through keen awareness. Being informed and paying attention to details is key.

The author's objectives in writing this story was to share his experience and offer a realistic look at what to expect with people who want to leverage technology to succeed in business.

For a more comprehensive guide to launching and managing an automation technology company refer to the book LEVERAGING TECHNOLOGY FOR SUCCESS. It contains a mix of facts and fiction (this story is part of the all-encompassing book).

Comments:

www.ingramcontent.com/pod-product-compliance
Lightning Source LLC
Chambersburg PA
CBHW071423180526
45170CB00001B/194